Computer Networks

Cluster Computing and Multi-Hop Network Research

COMPUTER NETWORKS

Topology Construction for Bootstrapping Peer-to-Peer Systems over Ad-Hoc Networks
Wei Ding
2009. ISBN: 978-1-60692-919-3

IP Packet Forwarding Research Progress
Pi-Chung Wang
2009. ISBN: 978-1-60741-016-4

File Sharing Applications Engineering
Luca Caviglione
2009. ISBN: 978-1-60741-594-7

Cluster Computing and Multi-Hop Network Research
Ciceron Jimenez and Maurice Ortego (Editors)
2010. ISBN: 978-1-60876-186-9

COMPUTER NETWORKS

CLUSTER COMPUTING AND MULTI-HOP NETWORK RESEARCH

CICERON JIMENEZ
AND
MAURICE ORTEGO
EDITORS

Nova Science Publishers, Inc.
New York

For permission to use material from this book please contact us:
Telephone 631-231-7269; Fax 631-231-8175
Web Site: http://www.novapublishers.com

LIBRARY OF CONGRESS CATALOGING-IN-PUBLICATION DATA

Cluster computing and multi-hop network research / editors, Ciceron Jimenez and Maurice Ortego.
p. cm.
Includes index.
ISBN 978-1-60876-186-9 (hardcover)
1. Parallel processing (Electronic computers) 2. Electronic data processing--Distributed processing. 3. Computer networks--Research. 4. Wireless communication systems--Research. I. Jimenez, Ciceron. II. Ortego, Maurice.
QA76.58.C5583 2009
004.6072--dc22
2009040812

Published by Nova Science Publishers, Inc. ✦ New York

CONTENTS

PREFACE

Processes in parallel applications running on high-performance computing (HPC) clusters communicate extensively with each other through the interconnection network. Therefore, the interconnection networks and their communication system software play a significant role in the performance of such clusters. This book studies a number of high-speed interconnects from different perspectives. It first evaluates the basic performance metrics of these interconnects, namely latency, bandwidth, and memory registration overheads. Then, some internal characteristics of their Message Passing Interface (MPI) implementations such as queue usage and buffer re-use effect. In addition, VANETS (Vehicular Ad-hoc Networks) are a form of MANETs (Mobile Ad-hoc Networks) that provide communication between vehicles and a fixed equipment or among nearby vehicles. This book focuses on the simulations of VANETs in the most possible realistic environments. Furthermore, spatial distribution models, like Poisson, binomial, and negative binomial functions have no power to predict spatial distribution. This book aims to develop some models, and use these models to describe or predict spatial distribution of arthropods. Other chapters in this book examine cluster analysis in a very general situation, in presence of mixed variables and replicates, present a linear pipeline scheme for image compositing in parallel visualization to support efficient image delivery to remote client, and provide a qualitative and quantitative comparison among the various protocols, from which future research studies are indicated.

Processes in parallel applications running on high-performance computing (HPC) clusters communicate extensively with each other through the interconnection network. Therefore, the interconnection networks and their communication system software play a significant role in the performance of such clusters. Ethernet, which has served as the dominant interconnect in wide area networks has shown to be inefficient for HPC applications. In this regard, several contemporary interconnects have been introduced that offer significantly lower latency and higher bandwidth compared to Ethernet. To take advantage of such interconnects in HPC clusters, there is a need to systematically assess and analyze their different characteristics at the user-level, and Message Passing Interface (MPI) layer. It is also important to understand how the communication subsystems of these interconnects interact with the emerging multi-core architectures, increasingly used in modern HPC cluster nodes. Chapter 1 studies a number of high-speed interconnects including InfiniBand, Myrinet-10G, and iWARP Ethernet from different perspectives. It first evaluates the basic performance metrics of these interconnects, namely latency, bandwidth, and memory registration overheads. The analysis is extended to multiple connection scalability of such interconnects to assess their ability to keep pace with growing multi-core platforms. The paper then analyzes some internal characteristics of their MPI implementations such as queue usage and buffer re-use effect. The network interface and the MPI ability to support overlap and communication progress is then studied. Results confirm that modern interconnects are able to effectively improve the performance of clusters by bypassing the operating system and reducing the CPU overhead.

VANETS (Vehicular Ad-hoc Networks) are a form of MANETs (Mobile Ad-hoc Networks) that provide communication between vehicles and a fixed equipment or among nearby vehicles. VANET communication is a big challenge due to mobility issues (mobility depends on drivers' plans but at the same time drivers react to changing road conditions). Specific protocols designed to support comfort and traffic safety applications should be developed; they must be able to adapt to mobile scenarios. Since it is very difficult to validate the implementation of these protocols in a real world environment, a common way of evaluation performance is through simulations.

In Chapter 2 the authors focus on the simulation of VANETs in the most possible realistic environments. It is necessary to integrate traffic and network simulation models because simulations based on simple mobility models do not reflect the reality of VANET scenarios. For this purpose, the most important current tools for the traffic simulation in a transportation environment are introduced. Afterwards, the open source micro-traffic simulator SUMO (Simulation of Urban Mobility), the popular tool for the simulation and evaluation of network protocols NS-2 (Network Simulator) and an efficient tool to rapidly generate realistic mobility models named MOVE (Mobility Model Generator for Vehicular Networks), have been selected. Realistic simulations in two typical VANET scenarios have been conducted: a highway and the streets of a city (one characteristic neighborhood of Barcelona (Eixample) in Spain). DYMO has been selected as a simple routing protocol that succeeds AODV and is very appropriate for mobile environments. Realistic simulations have been carried out with different number of vehicles, number of vehicles acting as sources, a variety of vehicles' speeds, several TCP and UDP flows, etc. The goodput and the impact of traffic collision in the proposed scenarios have also been analyzed.

Spatial distribution models, like Poisson, binomial, and negative binomial functions, etc., have no power to predict spatial distribution. Chapter 3 aimed to develop some models, and use these models to describe or predict spatial distribution of arthropods. Arthropod investigation was conducted on the grassland. An artificial neural network and a partial differential equation were developed for modeling field distribution of arthropods. Several models were compared and evaluated for their effectiveness in modeling spatial distribution.

Most arthropods on the grassland were insects. The most abundant orders were found to be Homoptera, Orthoptera, and Hymenoptera. The results demonstrated that artificial neural network exhibited an excellent simulation performance. Simulated spatial distribution was perfectly in accordant with the observed one. Overall neural network performed better in the case of lower total abundance on the grassland. Response surface model may better fit the spatial distribution of arthropods but the simulation performance of this model was lower than neural network. Cross validation revealed that neural network yielded much better performance than response surface model and spline function in predicting spatial distribution of arthropods. Confidence interval of modeled abundance could be obtained using randomized submission of quadrate sequences in the neural network modeling. A method for obtaining parametrical values in the partial differential equation was presented. The partial differential equation developed in this study may be used to make extrapolation on spatial distribution of arthropods. It is concluded that artificial neural network is a powerful model to describe the spatial distribution of arthropods.

The simulation of biomolecules like proteins in their natural environment is computational demanding and parallelization is a must to simulate relevant time scales within a reasonable computation time. The software COSMOS uses a mixed approach of classical molecular dynamics (MD) and quantum chemistry (QC) by incorporating atomic charges computed using the Bond Polarization Theory (BPT) into the force field. To parallelize the MD the authors used an atom-decomposition approach with replicated data. A kind of greedy algorithm yields a partition of the atoms without dividing chemical bonds. The parallel version of the force field COSMOS-NMR is implemented using MPI. In Chapter 4 the authors present the results of performance measurements. As test case the authors used an ice crystal heated over a picosecond. The computations are performed on a HPC cluster with InfiniBand. It turns out, that the authors code greatly benefits from the low latency of the interconnect.

In the biological and biomedical sciences, large multivariate datasets are frequently generated by investigators which include mixed data and replicate experiments. Recent technological advances, such as cDNA microarray technology have made possible the study of thousands of genes in a biological specimen. Given the flood of data that is generated from such experiments, appropriate analysis need to be performed so that useful knowledge and trends that would be otherwise undetectable can be observed. Chapter's 5 aim of interest is concerned with cluster analysis in a very general situation, in presence of mixed variables and replicates. To obtain several replicates is a straight forward method for improving the quality of inferences made from experimental studies. It is known that the replicates are not true duplicates, so the between-replicates variability must be taken in consideration. The usual practice is to average the replicate results before assessing differences between objects. This results in the loss of valuable information about variability. The authors describe a new partitioning cluster procedure based on a variant of Rao's diversity coefficient and in the so called path-distance. The approach uses top-down analysis. It starts with a large cluster and splits into smaller ones until each sub-cluster contains only one data point. The algorithm assigns each object to only one cluster, gives the global extreme (not local) of the function to be optimized and does not depend on the initial object selected to begin the algorithm with. The procedure takes into account the variance of the replicates. The methodology is a general algorithm that can be used to identify clusters in any situation, where data are arranged in tables (one for each replicate), where rows represent all objects for which data has been collected and columns represent the measured variables (binary, qualitative or continuous). The authors have tested the procedure on synthetic and on real data, obtaining good results and providing an easy-to-use bioinformatics solution for the clustering analysis of experiments with replicates. The program implementing the algorithm is available upon request from the authors.

The rapid advances in computing power made possible by the modern supercomputing technologies have enabled a wide spectrum of next-generation scientific applications with a primary goal of either simulating very complex physical phenomena or creating digital representations of real-world objects for educational or scientific purposes. These applications generate vast amounts of scientific data on the order of terabytes at present or petabytes in the predictable future, which must be transferred, visualized, and analyzed by a collaborative team of geographically distributed scientists. The capability of collaboratively analyzing and visualizing today's abundant scientific data and on-going computations over wide-area networks is critical to ensure both the success of mission-critical applications and the utilization of expensive computing resources. In Chapter 6, the authors discuss various issues in the design of scalable parallel algorithms for large-scale data analysis and visualization and present a number of novel techniques with focus on scalability improvement. The main goal of algorithm design for parallel computing is to balance workload as well as minimize runtime communication and synchronization costs. The authors present several data management and distribution schemes that promise highly scalable parallel performance by utilizing both out-of-core schemes and effective data traversal along a space filling curve. These schemes ensure balanced workload for different parallel implementations and avoid data replication and run-time redistribution to maximize the utilization of limited system resources. The authors also present a linear pipeline scheme for image compositing in parallel visualization to support efficient image delivery to a remote client. The linear pipeline scheme arranges an arbitrary number of parallel processors within a cluster in a linear order and divides the image into a carefully selected number of segments, which flow through the linear in-cluster pipeline and wide-area networks to the remote client consecutively. The authors analytically determine the segment size that minimizes the final image display time and derive the conditions where the proposed image compositing and delivery scheme outperforms the traditional scheme based on the binary swap algorithm. The efficacy of these approaches is demonstrated by case studies that involve data analysis and visualization techniques widely used in scientific fields.

Wireless mobile multihop networks, such as ad hoc and sensor networks, have become rather popular in realizing seamless distributed computation and accommodating autonomic behavior. In these networks, the lack of central infrastructure, the spatio-temporal dynamic nature of wireless communications along with their mobilityinduced dynamic topology, pose additional challenges on the traffic-carrying capacity of the system and raise concerns on the delay performance of packet delivery mechanisms. Despite such impediments of the wireless multihop environment, it has been recently shown that mobility can be exploited by various relay techniques for increasing the capacity of such networks. However, the throughput-effectiveness of the mobilityexploiting relay methods takes place at the cost of a greater number of intermediate hops traversed by packets traveling in the network in order to reach their final destination and additional buffering required at the intermediate nodes, revealing an inherent tradeoff between capacity and delay in mobile multihop networks. In Chapter 7, the authors first briefly review various mobility models, and then focus on the inherent mobilityinduced tradeoff between capacity and delay in multihop networks. Mobility is treated as both a constructive (increasing capacity), as well as a destructive (increasing delay) factor in wireless infrastructureless networks. Formal problem settings are presented, followed by applications and protocols that could be used in actual networks for increasing performance and realizing viable networking services. The authors also provide a qualitative and quantitative comparison among the various protocols, from which directions for future research studies are indicated.

In Chapter 8 the authors present an algorithm to approximate the surface of 2D or 3D objects combining concepts from geometric algebra and artificial neural networks. The authors approach is based on the self-organized neural network called Growing Neural Gas (GNG), incorporating versors of the geometric algebra in its neural units; such versors are the transformations that will be determined during the training stage, and then applied to a point to approximate the surface of the object. The authors also incorporate the information given by the generalized gradient vector flow to select automatically the input patterns, and also in the learning stage in order to improve the performance of the net. Several examples using medical images are presented, as well as images of automated visual inspection. Comparisons using snakes and a version of GSOM developed to work incorporating versors and gradient information, confirm that the authors approach is very promising. As a second application, it is shown a kind of morphing or registration procedure, namely the algorithm can be used when transforming one model at time t_1 into other at time t_2. The authors include also examples applying the same procedure, now extended to models based on spheres.

As explained in Chapter 9, the detection of stationary segments or zones in logging data sequences can be carried out by searching for change points in the data. These change points, which are associated to abrupt changes in the statistical properties of the underlying process, can be automatically identified by analyzing how the probability density function (pdf) of the data calculated in a short moving window varies with depth. Specifically, a change point is proposed when the difference between two consecutive pdf (or other statistical property) is significantly large. In a practical context, it is the statistical behavior of logging data (density, resistivity, porosity, etc.) that drives the zonation of a logging sequence into relatively homogeneous and stationary intervals, which are then associated to zones of similar lithologic character such as stratigraphic units and formations. When several logging data sequences are available, the segmentation can be carried out in a multivariate fashion. Here, all features of the individual sequences are captured simultaneously by the zonation algorithm. Naturally, this strategy leads to more reliable and consistent results because multiple data sequences are combined to produce a single sequence of segments. When applied to data from the Integrated Ocean Drilling Program (IODP), the authors show that the results of the multivariate zonation yields a high correlation with the available core-log lithology interpretation.

In: Cluster Computing and Multi-Hop Network Research ISBN: 978-1-60876-186-9
Editors: C. Jimenez and M. Ortego, pp. 1-34

Chapter 1

MODERN INTERCONNECTS FOR HIGH-PERFORMANCE COMPUTING CLUSTERS

Mohammad J. Rashti[a] and Ahmad Afsahi[b]
Department of Electrical and Computer Engineering,
Queen's University, Kingston, ON, Canada

Abstract

Processes in parallel applications running on high-performance computing (HPC) clusters communicate extensively with each other through the interconnection network. Therefore, the interconnection networks and their communication system software play a significant role in the performance of such clusters. Ethernet, which has served as the dominant interconnect in wide area networks has shown to be inefficient for HPC applications. In this regard, several contemporary interconnects have been introduced that offer significantly lower latency and higher bandwidth compared to Ethernet. To take advantage of such interconnects in HPC clusters, there is a need to systematically assess and analyze their different characteristics at the user-level, and Message Passing Interface (MPI) layer. It is also important to understand how the communication subsystems of these interconnects interact with the emerging multi-core architectures, increasingly used in modern HPC cluster nodes. This chapter studies a number of high-speed interconnects including InfiniBand, Myrinet-10G, and iWARP Ethernet from different perspectives. It first evaluates the basic performance metrics of these interconnects, namely latency, bandwidth, and memory registration overheads. The analysis is extended to multiple connection scalability of such interconnects to assess their ability to keep pace with growing multi-core platforms. The paper then analyzes some internal characteristics of their MPI implementations such as queue usage and buffer re-use effect. The network interface and the MPI ability to support overlap and communication progress is then studied. Results confirm that modern interconnects are able to effectively improve the performance of clusters by bypassing the operating system and reducing the CPU overhead.

Keywords: Cluster Interconnects, InfiniBand, iWARP Ethernet, Myrinet, Performance Analysis, MPI

---

[a] E-mail address: mohammad.rashti@queensu.ca
[b] E-mail address: ahmad.afsahi@queensu.ca

1. Introduction

Inter-process communication plays a significant role in the performance of computer clusters. Ethernet is the predominant protocol for wide area networks. However, the host-based implementation of TCP/IP over Ethernet is very expensive incurring high communication latency and high host CPU utilization. This has been magnified with the introduction of 10G Ethernet by increasing the load of communication processing on the host. None of these shortcomings is acceptable for today's CPU and memory hungry high-performance applications.

There have been two trends for modern interconnects to overcome Ethernet's inadequacies: cluster-specific interconnection networks with their advanced features, and accelerating communication stacks over Ethernet. Cluster-optimized high-performance interconnects such as Myrinet [31], InfiniBand [18], and Quadrics[3] with fast communication processing engines have been introduced, making clusters the predominant platform of choice for high-performance computing. Such interconnects are the backbone for scalable, high-performance clusters by offering very low latency and high bandwidth communication. These interconnects employ modern techniques such as Operating System (OS) kernel bypass with user-level libraries, protocol processing offload to release main CPU cycles, and *Remote Direct Memory Access* (RDMA) [14]. RDMA acts as a one-sided operation that allows access to the memory of a remote node. It allows direct data transfer from the source buffer to the destination buffer at the remote node without the host processor intervention or any intermediate copy. This inherent zero-copy feature of RDMA, in concert with the OS bypass mechanism, effectively helps save CPU cycles and memory bandwidth that are partly responsible for host CPU overhead. RDMA has been exploited to improve the performance of both one-sided and two-sided communications in message passing over InfiniBand [19][48] as well as Quadrics QsNet^{II} [3], to name a few. It also forms the skeleton of communication in Myrinet's messaging layer.

In accelerating Ethernet-based communication processing, *stateless offload* and *TCP/IP Offload Engines* (TOEs) [13] have emerged as solutions to offload protocol processing from the host CPU to the hardware on the Ethernet *Network Interface Card* (NIC). In addition, RDMA has been utilized for Ethernet-based stacks to potentially keep Ethernet as a viable network for HPC and data center applications. RDMA consortium [42] and IETF [16] have developed a set of standard extensions to TCP/IP and Ethernet to eliminate the host CPU overhead. These specifications are collectively known as iWARP. The specification proposes a set of descriptive interfaces, called iWARP verbs, allowing direct access to the *RDMA enabled NIC* (RNIC), thereby bringing the benefits that a proprietary interconnect can offer.

MPI is a message passing standard proposed in 1993 by MPI forum for loosely-coupled parallel machines [29]. The standard provides a set of portable messaging interfaces that could be implemented on any machine. Each MPI implementation may add certain features that are not normally required by the standard. Currently, there are two MPI specifications: MPI-1 and MPI-2. MPI-2 was introduced in 1997. Since MPI is now the de facto standard for message passing we will use it as the main communication standard in our evaluation of modern interconnects. MPI Forum is currently making revisions to the standard and finalizing the next generation of MPI known as MPI-3.

We have done an extensive performance evaluation of the communication architectures of contemporary interconnects in order to signify their contributions to high-performance computing. Our evaluations are at both the user-level and MPI layer. Our benchmarks at the user-level include the basic latency and bandwidth for mainstream communication primitives, multiple-connection latency and bandwidth, and cost of memory registration. At the MPI level, we present and analyze the basic latency and bandwidth, multi-pair latency and bandwidth, MPI queue usage and buffer reuse effect, communication/computation overlap and communication progress ability of modern interconnects.

The basic results confirm that the modern high-performance interconnects are expectedly performing much better than the regular Ethernet. For example, our results show that using the same hardware, a 10G Ethernet-based MPI is about eight times slower than an InfiniBand-based MPI implementation. The bandwidth is also highly utilized when using modern user-level libraries and offloaded protocol processing. For instance, InfiniBand and iWARP Ethernet utilize more than 90% of their maximum available bandwidth. This has been achieved through advanced features supported in modern interconnects such as OS bypass with user-level libraries, offloaded communication processing, skipping unnecessary data copies inside the host, and the ability to directly access the remote memory using RDMA.

The rest of this chapter is organized as follows. Section 2 provides an overview of the modern interconnects studied in this chapter. Related work is reviewed in Section 3. Section 4 describes our experimental platform. Basic user-level and MPI performance results are presented in Section 5. In Section 6 through Section 9, we evaluate in-depth characteristics of the interconnects including multiple-connection scalability, memory registration and buffer-reuse effect, MPI internal queue performance, communication/computation overlap, and communication progress ability. Finally, in Section 10 we conclude the chapter.

2. Overview of the Modern High-Performance Interconnects

2.1. Mellanox Infiniband

InfiniBand (IB) is an I/O interconnection technology consisting of end nodes and switches managed by a central subnet-manager [18]. End nodes use Host Channel Adapters (HCA) to connect to the network. InfiniBand verbs form the lowest level of software to access the IB protocol-processing engine offloaded to the HCA. IB verbs support four transport models for communication: *Reliable Connection* (RC): connection-oriented service with guarantied delivery; *Unreliable Connection* (UC): connection-oriented but unreliable service; *Reliable Datagram* (RD): connection-less service with guarantied delivery - not available in the current hardware; and *Unreliable Datagram* (UD): connection-less and unreliable service [18]. The verbs layer has queue-pair (QP) based semantics, in which processes post send or receive work requests (WR) to send or receive queues, respectively. Queue-pairs are able to work in any of the above transport models. The WR is used by the user to send data from the source process address space into the send queue, or to wait for a matching data to arrive into the receive queue. A completion queue associated with the QP is used by the hardware to place completion notifications for each WR. IB verbs require registration of memory buffers prior to their use.

InfiniBand supports two models of communication semantics: channel semantics, supported by traditional two-sided send and receive operations in which a receive request is needed to be pre-posted before a send message arrives. This is to make sure that the arrived message can be directly placed in the remote memory while the memory address does not need to be exposed remotely. Memory semantics, using one-sided RDMA operations, is used when the remote memory address is exposed and can be directly accessed without the intervention of the local host. Connection-oriented communication models support memory semantics, while connection-less models have limited or no support for it.

ConnectX [7] is the most recent generation of IB HCAs, released by Mellanox Technologies [26]. In addition to standard IB features, ConnectX cards support a number of additional functionalities, including a memory-friendly alternative to IB reliable connection transport called *eXtended Reliable Connection* (XRC), reliable multicast capability and extended quality of service support [7]. The XRC transport service is an extension of the Reliable Connection transport service [22]. Along with the Shared Receive Queue, this transport service enables mitigating the number of transport connections required to be established by applications, by specifying/controlling a receiver-side shared receive queue by the senders. ConnectX also allows combination of XRC with shared send queues, which reduces the required resources even more. Using the *Virtual Protocol Interconnect* (VPI) [50], dual-port ConnectX cards can also be programmed to operate simultaneously in both 10G Ethernet and IB modes. This helps to seamlessly operate a ConnectX card on either or both Ethernet and IB networks at the same time.

2.2. Myricom Myri-10G

The most recent Myricom product is the Myri-10G networks accompanied with 10G Myrinet Express (MX-10G) user-level library [31]. The Myri-10G NICs are dual-protocol cards that support 10G Ethernet and 10G Myrinet. Myricom has its own Myri-10G switches, but the NICs can also be used in an Ethernet environment using standard 10G Ethernet switches. Moreover, recent releases of MX-10G libraries can be used over Ethernet network, called MX over Ethernet (MXoE). In addition to Myrinet network protocol processing, Myri-10G NICs offload some stateless network processing such as checksum calculations, segmentation and re-assembly, but they are not full TCP Offload Engines (TOE).

The MX-10G library has semantics close to MPI. Basic communication primitives are non-blocking send and receive operations that are directly used in the implementation of MPI. The library registers user buffers internally. The MX library pipelines messages on the wire for increased bandwidth. The MX library uses three communication channels: the network channel where messages are transferred through the NIC and the network; the shared-memory channel where messages are exchanged between processes on the same machine using shared-memory, and the self channel for intra-process message that is optimized for messages sent by a process to itself [31]. The use of these channels for different communication modes (i.e. inter-node, intra-node, self) is regulated by some environment variables. However, by default, intra-node communication is done using the shared-memory channel.

2.3. 10G iWarp Ethernet

RDMA consortium [42] has developed the iWARP standard as an extension to TCP/IP to ensure low overhead user-level communication over offloaded TCP/IP stack. The specification proposes a set of descriptive interfaces at the top layer, called *iWARP verbs* [14]. Verbs provide a user level abstract interface for direct access to the RNIC, which provides more benefits than a simple TOE by offering direct data transfer ability and OS bypass using RDMA. Verbs have QP based semantics similar to InfiniBand, and require user buffers to be locked down before the data transfer takes place. The iWARP standard has a multi-layer protocol stack. The verbs layer is at the top layer of software. High-level RDMA-based sockets, User-level Direct Access Transport API (uDAPL) [11] and MPI are implemented on top of the verbs layer and help implement portable user applications over different RNICs.

Below the verbs layer, there is the *RDMA Protocol* (RDMAP) layer that is responsible for performing RDMA operations and supplying communication primitives for verbs layer remote memory access calls. The data transfer primitives are Send, RDMA Write and RDMA Read operations [14]. Upper layer requests are delivered in order from RDMAP to the next layer. The Send and RDMA Write operations require a single message to be exchanged, while the RDMA Read needs a request by data sink, followed by response from the data source.

The *Direct Data Placement* (DDP) layer is used for direct transfer of data from the user buffer to the iWARP RNIC. DDP supports two types of data placements: tagged and untagged. For the tagged model used for RDMA (one-sided) operations, the operation source provides a reference to the data buffer address (sender-based buffer management), while in the untagged model used for Send (two-sided) operations, the operation target specifies the data buffer address (receiver-based buffer management) [44]. An intrinsic overhead for RDMA operations is that a previously advertised (tagged) data buffer at the target needs to be sent to the source prior to the actual operation.

The *Marker PDU Aligned* (MPA) layer is used to put boundaries on DDP messages transferred over stream oriented TCP protocol [8]. MPA inserts markers into protocol data units prior to passing them to the TCP layer. It also assembles marked data units from TCP stream and removes the markers before passing them up to the DDP layer. Obviously, due to the message-oriented nature of SCTP protocol [17], there is no need for an MPA layer when SCTP is used instead of TCP.

3. Related Work

Due to the contributions that cluster interconnects are meant to make to high-performance applications, it is essential to assess the features and performance of these networks in a systematic fashion. In this regard, researchers have studied various interconnection networks at different layers. The work in this chapter is along the work of other researchers in the systematic assessment of high-performance interconnects for modern clusters. However, this chapter contributes by providing an extensive study of the state-of-the-art interconnects over emerging multi-core clusters at both the user-level and MPI level.

A large body of work exists on the performance analysis of modern cluster-specific interconnects [4] [25] [24] [5] [38] [51] [15]. In [4], the authors compare Cray 3E, IBM SP, Quadrics, Myrinet 2000 and Gigabit Ethernet. They measure the LogP [20] parameters of all interconnects and show that in some cases the overlapping of sender and receiver overheads causes the LogP assumptions to be inaccurate.

The authors in [25] compare the performance of previous generation of InfiniBand, Myrinet and Quadrics at the user-level (using VAPI, GM and Elan). In [24], an MPI level comparison of IB, Myrinet, and Quadrics networks is presented. The paper presents the MPI latency and bandwidth, overlap, buffer re-use, and memory consumption. The authors show the scalability of these networks using some benchmarks.

A comparison of the InfiniBand network with Quadrics Elan-4 has been reported in [5]. The authors compare the two networks in terms of memory registration, independent progress, offload and overlap capability. They also present the latency and bandwidth at the MPI level and user level. The paper also contains the results for scaling efficiency using LAMMPS benchmark [46]. They finalize the paper with an interesting cost comparison between Quadrics and IB.

A comprehensive performance analysis of Myrinet 2000 is presented in [38]. The paper presents the latency and bandwidth results at the GM level as well as MPICH over GM. The authors discuss buffer re-use, overlap, LogP and collective communication results as well as bandwidth for uniform and permutation traffics. The same authors perform a study on two port Myrinet networks with GM2 library in [51]. The paper includes some non-data-transfer tests such as memory registration and deregistration costs. It also includes results for GM RDMA performance, MPI level benchmarks for latency and bandwidth, overlap, impact of polling or blocking, and collective communication [51]. The authors characterize some NAS parallel benchmarks in terms of their communication characteristics and present their performance on the Myrinet network.

In [15], performance of MPI over 10G Ethernet has been analyzed. In addition to basic MPI micro-benchmark tests such as latency and throughput, the authors report the amount of MPI overhead over TCP/IP library. Their results show that moving toward 10G Ethernet does not benefit MPI applications as much as it benefits TCP/IP based applications, primarily because MPI communication semantics is not in match with that of TCP/IP. The authors conclude the significance of recent revolutionary changes in Ethernet adapters (using TOE or RDMA over IP) [15] to fulfill MPI performance requirements.

A few works have been reported on Ammasso RNIC, an early iWARP enabled Gigabit Ethernet adapter released in 2004 [9] [21] [43]. In [9], the authors reported some basic performance results for the Ammasso RNIC including comparison of MPI latency and bandwidth with those of TCP and iWARP. The paper also presents non-data transfer operation costs such as the costs of queue-pair operations and memory registration.

In [21], the authors compare the communication latency over Ammasso iWARP verbs and TCP sockets for wide area network applications. They show that RDMA is beneficial especially under heavy load conditions. In fact, RDMA can provide better communication progress and requires less CPU resource compared to the traditional sockets over TCP/IP. Another work in [43] showed that RDMA Ethernet over Ammasso achieves significant performance improvement over conventional Ethernet, with some applications approaching the performance of IB.

A few works also exist on the first generation of 10G iWARP adapters. In [10], some preliminary performance results is reported comparing the NetEffect NE010 RNIC [33] on a 133MHz PCI-X bus with a Mellanox 4X IB card on an x8 PCI-Express bus. In [39] a more extended analysis of the NetEffect NE010 RNIC for high-performance computing has been presented. In [39] the NetEffect verbs layer and its MPI implementation is investigated in comparison to InfiniBand and Myrinet, with various micro-benchmarks for different communication performance abilities. Their results highlight the contribution of RDMA enabled TOEs to the Ethernet performance.

In [32], the authors present the design, implementation and evaluation of the first MPI implementation for iWARP over OpenFabrics verbs [35]. Their iWARP implementation over Chelsio 10G iWARP RNICs offers less than 7μs MPI latency and more than 1200MB/s MPI bandwidth on top of RDMA operations. The paper also presents performance for MPI one-sided operations (Put and Get), and some collective operations.

A small body of work exists concerning the analysis of overlap and communication progress in contemporary networks [6] [5][40] [12] [48] [47] [45]. In [6], the authors discussed the concept of independent progress, offload and overlap, and compared the impact of six MPI implementations on application performance running on two platforms with Quadrics QsNet [36] and CNIC network interface cards (used in ASCI Red machine) [6]. Their results show that in almost all benchmarks, the combination of offload, overlap and independent progress significantly contributes to the performance. The study in [5] concerns a similar work on InfiniBand and Quadrics QsNet^{II}. The work in [40] is along the same direction as in [6] [5] with the difference that it is focused at analyzing the ability of modern interconnects and their MPI implementations in supporting communication progress and overlap rather than an application's ability in leveraging them. While [24] [51] address combined send and receive overlap, our proposed overlap measurement method in this study, along with [12] [47] [48] [40] target non-blocking send and receive overlaps separately.

Among the work on measuring the overlap ability of networks, the authors in [12] have proposed an overlap measurement method for MPI. They first compute the communication overhead, and then application availability is calculated using the overhead amount. In [40], overlap and progress ability of modern interconnects and their MPI implementations are measured by directly measuring the amount of computation that can be overlapped with communication.

The researchers in [47] have presented an instrumentation framework to estimate the lower and upper bounds on the achieved overlap for both two-sided and one-sided communication over IB. In [45], a method is proposed for quantifying the potential overlap of MPI applications. Considering a producer-consumer model of communication, the authors define the potential overlap in an application as the time between production of a datum at the producer (sender) and its consumption at the consumer (receiver). Experimental results are also presented for some real applications [45].

On improving communication progress and overlap, researchers have proposed RDMA Read based Rendezvous protocols for MPI, with the asynchronous progress ability. They have achieved very high overlap and up to 50% progress improvement relative to the RDMA Write Rendezvous protocol [48]. Some improvements have been made to the RDMA Read based Rendezvous protocol, aimed at improving its overlap and progress ability and/or decrease its asynchronous overhead [23] [41].

4. Experimental Framework

We have conducted our experiments using Dell PowerEdge R805 servers. Each machine is a dual-processor (SMP) machine with two AMD Quad-core Opteron 2.0GHz processors. Each core has a 512KB dedicated L2-cache and there is an 8MB L3 shared cache on each processor die. Each server has 8GB total physical memory and three x8 PCIe slots. The machines run Linux Fedora Core 5 SMP for x86_64 architecture with kernel 2.6.20.

Our iWARP network consists of the second-generation NetEffect two-port NE020 10G Ethernet RNICs [34], each with a PCIe x8 interface and CX-4 board connectivity. We use OpenFabrics verbs in OpenFabrics Enterprise distribution (OFED) v1.4 [35] to interface the cards. A Fujitsu XG700-CX4 12-port 10G Ethernet switch is used to connect the cards together. We have used the single-port Myri-10G NICs with 10Gase-CX4 ports [31], each with a PCIe x8 interface. Myri-10G NICs are operating in MXoE mode using MX-10G version 1.2.8 and are connected to our Fujitsu 10G Ethernet switch.

Our InfiniBand network consists of DDR (20Gb/s) Mellanox ConnectX two-port HCA cards [7], each with a PCI-Express (PCIe) x8 interface, connected to a Mellanox 24-port MT47396 Infiniscale-III switch [26]. The OpenFabrics verbs in OFED 1.4 are used as the user-level software interface for ConnectX. We have programmed the ConnectX cards in VPI mode in a way that one port could operate in IB while the other port is in 10G Ethernet mode. We run our 10G Ethernet tests using the 10G Ethernet port of the ConnectX cards, but we never use the IB and Ethernet ports at the same time.

We use MVAPICH 1.1 and MVAPICH2 1.2p1 [30] supplied in OFED distribution for IB and iWARP networks, respectively. For Myri-10G, we use MPICH-MXoE based on MPICH 1.2.7..1 [2]. MPICH2 1.0.8 [1] with Sockets and Shared memory (SSM) channel is used for tests over 10G Ethernet network. MVAPICH2 over iWARP switches from Eager to Rendezvous protocol at 32KB. This threshold for MPI on top of Myrinet and InfiniBand is set to 32KB and 9KB, respectively. Both MVAPICH and MVAPICH2 implement the Rendezvous protocol using RDMA Write and RDMA Read operations [48]. Unless explicitly specified, we are using the default RDMA Write based Rendezvous protocol in these MPI implementations.

In the following, we will be presenting and analyzing the experimental results concerning the performance of modern interconnects. In our experiments, we use user-level and MPI micro-benchmarks. For each test, we report the performance results averaged over 1000 measurements.

5. Baseline Performance

In this section, we investigate the basic performance abilities of modern interconnects, namely their small message latency and large data transfer bandwidth. Section 5.1 presents the user-level latency and bandwidth results, and Section 5.2 discusses the MPI results.

5.1. User-Level Latency and Bandwidth

We present the latency and bandwidth of two user-level communication libraries over three interconnects: OpenFabrics verbs used for InfiniBand and iWARP networks and MX-10G over Ethernet (MXoE) for Myrinet cards. Unlike the MX-10G library, where communication is based on non-blocking Send/Receive operations, tagged RDMA operations with QP-based semantics form the main communication model in the IB and iWARP protocol stacks. For RDMA operations, the remote memory address tag needs to be exchanged prior to any data transfer. The communication model is one-sided; therefore, an explicit synchronization is required at the end of communication to make sure the data are transferred completely. This kind of synchronization can be done using a Send/Receive (untagged) operation. However, to measure optimistic results, the completion of the RDMA write operations is checked by polling the target buffer.

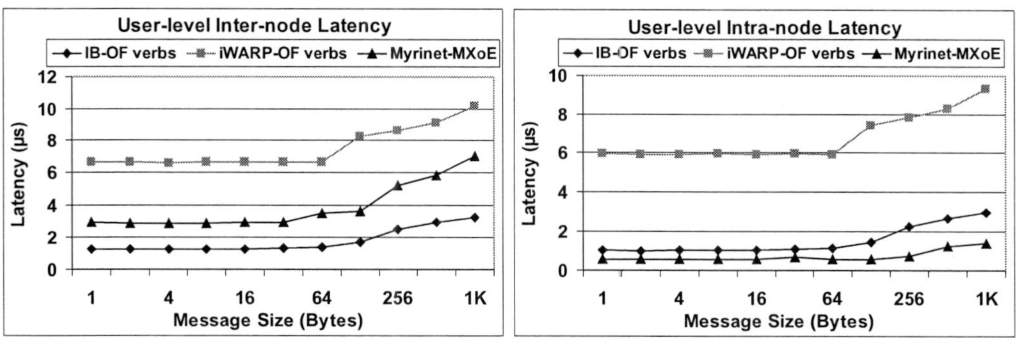

Figure 1. User-level latency.

Figure 1 presents the intra-node and inter-node latency of the user-level libraries. For MX-10G, we use the MX standard *mx_pingpong* latency and *mx_stream* both-way bandwidth tests. For iWARP and IB latency tests, we report the results from standard *ib_rdma_lat* latency and *ib_rdma_bw* both-way bandwidth tests. The user-level small message latencies for iWARP, InfiniBand and Myrinet are 6.6μs, 1.2μs and 2.9 μs, respectively. The intra-node latencies for IB and iWARP are not very different from the inter-node latencies, because the data travel over the NIC hardware and then loop back into the same node instead of going over the wire. Therefore, it is essentially bypassing the wire and switch latencies. For the Myrinet, however, this is not the case. By default, the MX library uses shared-memory channel for the intra-node communication.

The bandwidth results in Figure 2 show that while both InfiniBand and iWARP networks are very close to saturation (90%) for message around 4KB and larger, Myrinet is slowly saturated. The jump in Myrinet's bandwidth for 64KB messages is due to switching to a receiver-based Rendezvous protocol that directly transfers the data to the receiver, using a direct GET (RDMA Read) operation. The maximum both-way user-level bandwidth for iWARP is 2200MB/s, which represents 89% of its maximum available bandwidth (2.5GB/s). The bandwidth of Myrinet gets closer to the available 2.5GB/s by providing more than 2300MB/s bandwidth. On the other hand, the IB verbs saturate 75% of the maximum IB DDR bandwidth, roughly 3100MB/s. The limiting factor here is the PCI-Express 1.0 maximum throughput for un-encoded data which is 3200MB/s.

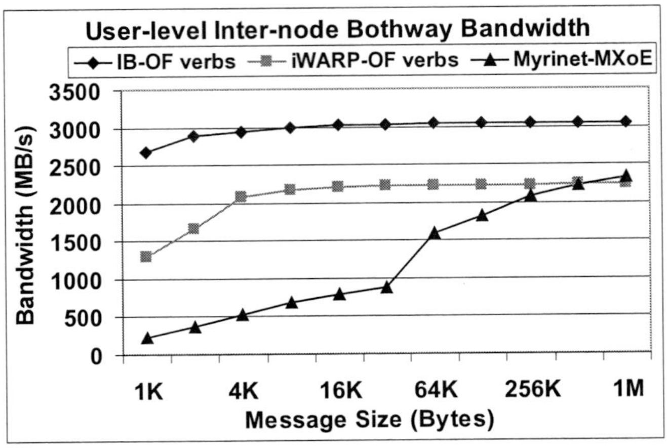

Figure 2. User-level bandwidth.

5.2. MPI Latency and Bandwidth

In this section, we present the MPI latency and bandwidth results. In our tests, we bind the affinity of processes to processors in order to avoid the overhead of process migration on cache performance.

5.2.1. MPI Latency

We measure the latency using a standard ping-pong communication method, in which the latency is measured as half round-trip time of a message between two processes. Figure 3(a) illustrates the inter-node MPI ping-pong latency of the interconnects, for small messages. The small message latency is around 7μs for iWARP, 1.5μs for MVAPICH-IB, and 3μs for MPICH-MXoE. Compared to Ethernet with the minimum latency of around 13μs over VPI ConnectX, one can observe that there is a significant improvement in small message latency for modern cluster interconnects, specifically the InfiniBand network.

Figure 3(b) shows the latency overhead of the MPI implementations over their respective user-level libraries. Results show that MPICH-MXoE offers the lowest overhead among the interconnects. This is because the MX-10G is the only library with communication semantics close to that of MPI. High overhead for some small messages in MVAPICH-IB shows the high overhead of buffer copies, compared to the small latency of data transfer over the network. Part of the overhead is also due to semantics miscorrelation between MPI and the user-level library. The reason for peaks in both MVAPICH-IB and MVAPICH2-iWARP overhead curves is due to switching from Programmed IO (PIO) based communication to DMA-based communication. In IB, the card uses DMA for 128 bytes or larger messages. This happens at 64 bytes for iWARP. The reason that MPI overhead for these message sizes is high is that for MPI tests the MPI header is added to the user data, making the actual transferred data larger than the PIO limit. Thus, the user-level library is transferring the data using PIO, while the same data with MPI header needs to go over DMA.

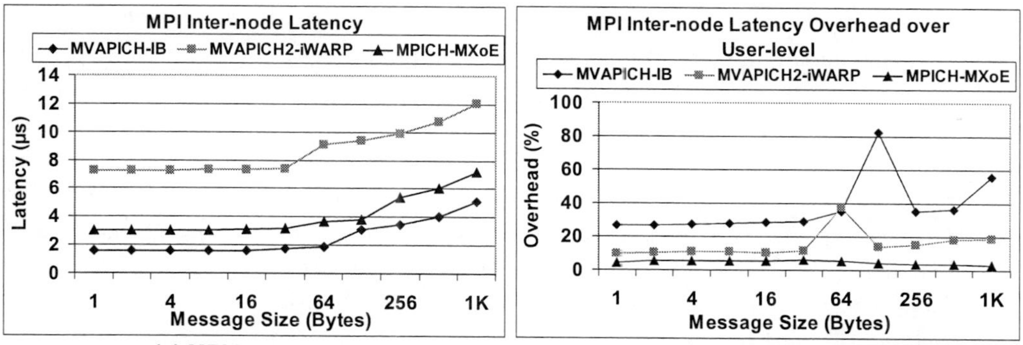

(a) MPI latency (b) MPI overhead over user-level

Figure 3. MPI ping-pong latency and its overhead over user-level.

Figure 4 shows the MPI intra-node latency for small messages. The inter-CMP latencies are measured between processes on different processor chips, while intra-CMP values are for processes on the same processor (CMP stands for Chip Multi-Processor). Very low inter/intra-CMP latency for small messages confirms that the MPI implementations bypass the network for intra-node communications. Comparing the intra-CMP to inter-CMP communication results, one can confirm that the message latency between two cores on the same processor chip is up to 30% lower than the latency between two separate processor chips. This is because of L3 cache sharing among cores on the same chip.

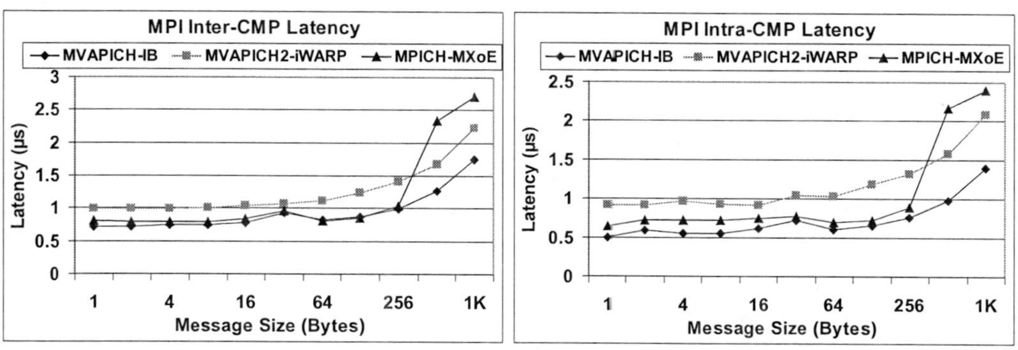

(a) Inter-CMP latency (b) Intra-CMP latency

Figure 4. MPI intra-node latency.

5.2.2. MPI Bandwidth

The other characteristic of modern cluster interconnects is their ability to transfer large amounts of data across the network with a high bandwidth due to avoiding kernel and CPU involvement and using low overhead communication architecture and protocols. In this section, we show how using these interconnects can improve the achieved communication bandwidth. In addition to reporting the ping-pong bandwidth, we measure the MPI inter-node bandwidth using a standard both-way communication method. The results for ping-pong and both-way bandwidth benchmarks from 1KB to 1MB message sizes are shown in Figure 5. In the ping-pong test, up to 1100MB/s for iWARP, 1200MB/s for Myrinet-10G, and 1600MB/s for MVAPICH-IB can be achieved. For all of the three networks, this is very close to (more than 85% of) the maximum one-way bandwidth on our platform.

In the both-way test, both the sender and receiver post a window of non-blocking send operations followed by a window of non-blocking receive calls. This puts more pressure on the communication and I/O subsystems. While a maximum bandwidth of 2300MB/s is achievable with the NetEffect iWARP and Myricom's Myri-10G cards, representing 92% network utilization, MVAPICH-IB is nearly saturating the maximum chipset bandwidth of around 3200MB/s. MPI over Ethernet bandwidth results (not shown) are notably worse than modern interconnects. Reaching 1400MB/s at 64KB in both-way bandwidth test, 10G Ethernet shows 55% utilization of its available bandwidth.

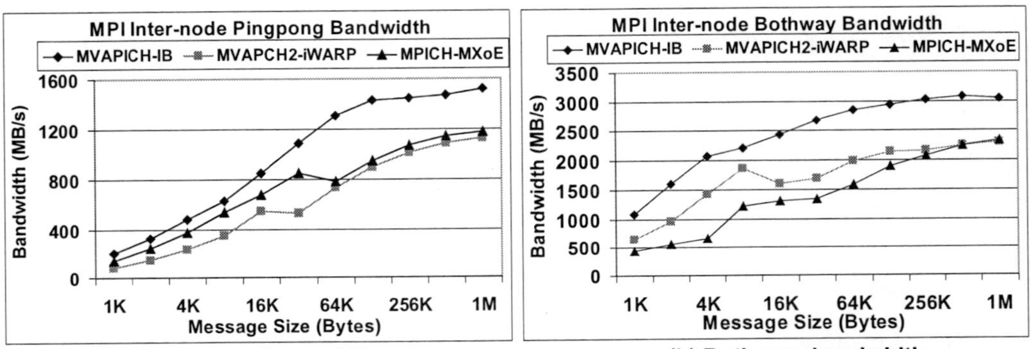

(a) Ping-pong bandwidth **(b) Both-way bandwidth**

Figure 5. MPI inter-node bandwidth.

6. Multiple-Connection Scalability

Today, multi-core multiprocessor nodes form the dominant building blocks of high-performance clusters. In such clusters, each core will run at least one process with connections to several other processes. Therefore, it is very important for the NIC hardware and its communication software to provide scalable performance with the increasing number of connections.

Both the iWARP and InfiniBand standards have a QP based communication model with similar semantics. Both support connection-oriented RDMA operations. Myrinet has an analogous semantics using MX endpoints that work similarly to connections in InfiniBand and iWARP standards. Therefore, it is worth showing how the hardware implementation of these interconnects and their MPI libraries support multi-connection scalability. In this section, we evaluate the multiple-connection scalability of the networks at two levels. We use OpenFabrics and MX-10G libraries at the user-level, and MPI non-blocking send and receive operations at the MPI level.

6.1. Multiple Connections at the User-Level

For each network, we pre-establish a number of connections (up to 16 connections) between two processes running on two nodes (each node has only one NIC) and then start communicating over those connections at the user-level. We measure the normalized round-trip latency and aggregate throughput over multiple connections. The results show how well communications over multiple connections can be performed in parallel.

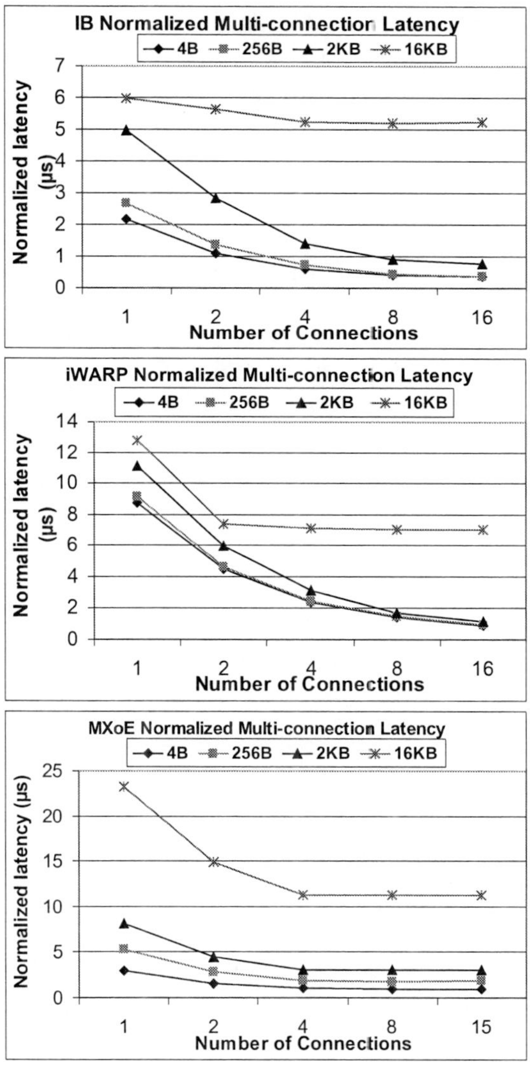

Figure 6. User-level normalized multiple-connection latency.

To determine the latencies, we performed a ping-pong test using all of the connections in parallel. A fixed number of messages are sent and received over the connections in a round-robin fashion. We vary the number of connections and message sizes and report the cumulative half round trip time divided by the number of connections and messages as the normalized multiple-connection latency. For the throughput test, a both-way communication is performed, where each process sends messages to its peer in a round-robin fashion over the established connections. The test lasts for a certain amount of time and at the end, the throughput is reported as the ratio of the amount of data transferred over the communication time.

As shown in Figure 6, by increasing the number of connections, the normalized multiple-connection latency of small messages for both iWARP and IB cards decreases even for 16 connections. This reflects the ability of the cards to keep up with a large number of parallel connections for small messages. A relatively fixed latency for messages larger than 2KB on these cards shows a serialization of communications over multiple connections. The story for Myrinet is a bit different. For Myrinet, for all message sizes, the normalized latency decreases well, but only up to four connections (endpoints). This clearly indicates that for all message sizes, Myrinet cards are able to keep up with only up to four connections. Note that the Myrinet results are reported up to 15 active connections, as the MX software on Myrinet cards does not allow more than 15 endpoints.

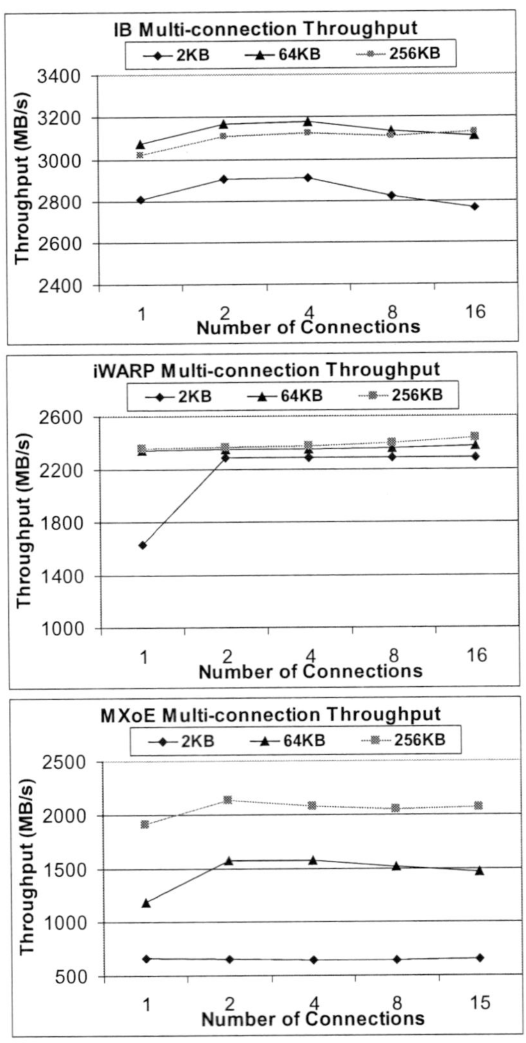

Figure 7. User-level multiple-connection aggregate throughput.

Figure 7 shows the aggregate multiple-connection throughput results. We report the results for messages up to 256KB. Obviously, all of the network cards are able to sustain a high level of bandwidth even when multiple connections are actively sending data over the cards. The striking observation is for the Myrinet where the throughput does not scale well with the message size, even with a high number of connections. It only scales slightly up to two connections. For example, with 64KB messages, while InfiniBand and iWARP almost saturate the link over a single connection, Myrinet reaches no more than two third of its link capacity, even with the highest number of active endpoints.

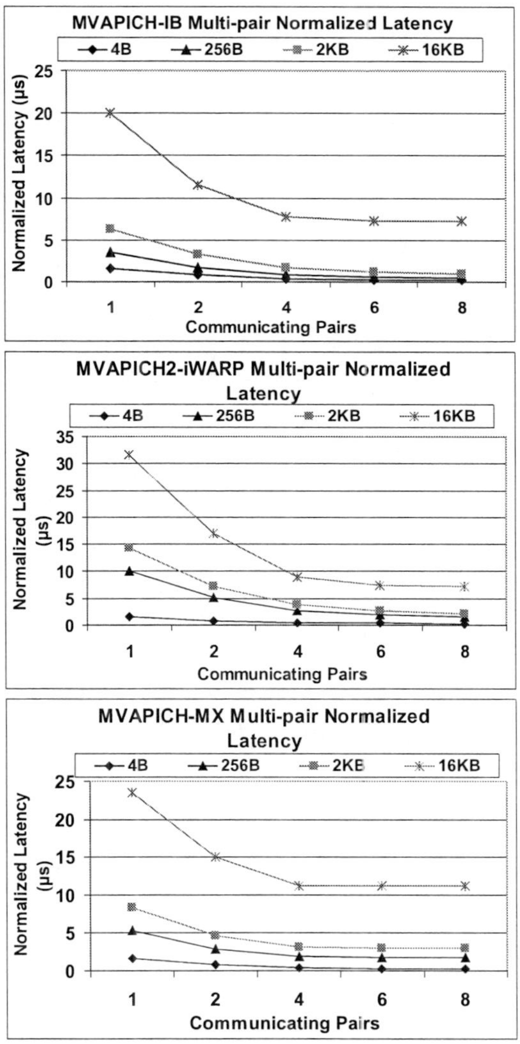

Figure 8. MPI-level normalized multi-pair latency.

6.2. Multiple Connections at the MPI Level

The preceding results were gathered at the user-level. In this section, we present the multi-connection scalability results at the MPI level. Similarly, two micro-benchmarks are used, one for measuring the normalized multi-connection latency and the other for the aggregate throughput. We use multiple pairs of MPI processes where the two processes in each pair communicate with each other. Due to the connection-oriented implementation of MPI libraries, each pair represents a connection over the NIC. For the latency test, we perform a ping-pong test and measure the average half-way round-trip time divided by the number of pairs. For the throughput test, a both-way communication test is performed and the throughout among multiple pairs is accumulated. Since our machines have eight cores per node, we can only experiment with up to eight concurrent communicating pairs.

In Figure 8, we report the average normalized latency over multiple pairs for all three interconnects. Similar to what we observed in Figure 6 at the user-level, there is a decreasing trend in normalized latency, showing some level of scalability with growing number of active connections on a multi-core platform. Regardless of message size, this scalability stops after four pairs for Myrinet, while for InfiniBand and iWARP we still see some decrease in the normalized latency even for eight pairs, especially for small messages.

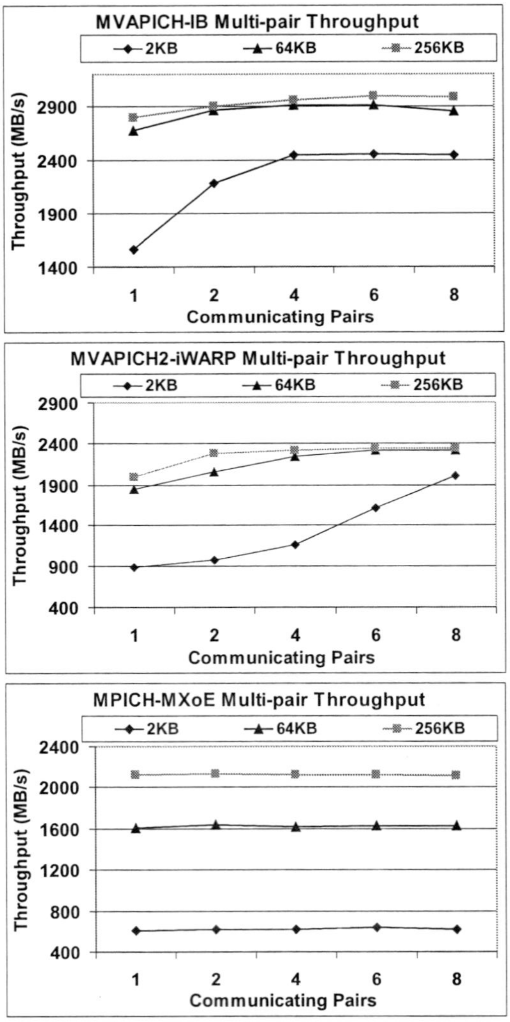

Figure 9. MPI-level aggregate multi-pair throughput.

Figure 9 shows the achieved aggregate bandwidth with different number of communicating pairs. In these tests, each communicating pair performs the both-way bandwidth test. Then, the individual measured bandwidths are added up as the aggregate bandwidth. To ensure all pairs start communication simultaneously, all processes initially participate in a global barrier synchronization. Clearly, when increasing the number of communicating pairs, iWARP and IB bandwidth increases very fast, regardless of the message size. On the other hand, similar to the user-level results, Myrinet shows a slow scale-up trend. Its throughput for a specific message size is almost fixed, regardless of the number of pairs. This implies that the card is carrying data from all pairs into a single communication-processing path. In other words, the communication processing appears to be serialized, not being able to saturate the available bandwidth even with multiple pairs and using relatively large messages. Comparing the user-level and MPI results shows a notable similarity between them. This indicates that the results reflect the characteristics of the network and their user-level library in handling multiple connections, even if the connections are handled by a single process pair.

7. Memory Registration and Buffer Reuse Effect

Reusing application buffers for communication is a method that can help avoid a number of buffer management related overheads. Cache and TLB misses are the common sources of overhead. One of the most important overheads in RDMA-based networks is the memory registration/deregistration. In this section, we intend to assess the overhead of memory registration in addition to the effect of buffer reuse in RDMA-based interconnects.

7.1. Memory Registration Effect

Memory registration is an expensive process that involves buffer pin-down and virtual-physical address translation [27]. In addition, the registration tag needs to be advertised to the remote node. This is why in the MPI Eager protocol the small messages are copied from application buffers into pre-registered and pre-advertised intermediate buffers. For the MPI Rendezvous protocol though, the application buffers are registered to be directly used for zero-copy transfer. The iWARP and InfiniBand standards require explicit registration of memory buffers prior to using them for communication. However, the Myrinet MX library does not require the user to register/de-register the memory. Instead, MX library internally performs the registration and de-registration.

Figure 10 depicts the memory registration and deregistration costs in IB and iWARP for different buffer sizes. The results clearly show that memory registration is costlier than memory deregistration for iWARP. In fact, using iWARP RNICs, memory deregistration cost is almost independent of the buffer size except for large buffers. However, for InfiniBand, deregistration is a bit more expensive for small buffers but the trend is similar to iWARP for large buffers, for which registering a buffer is much more expensive than deregistering it.

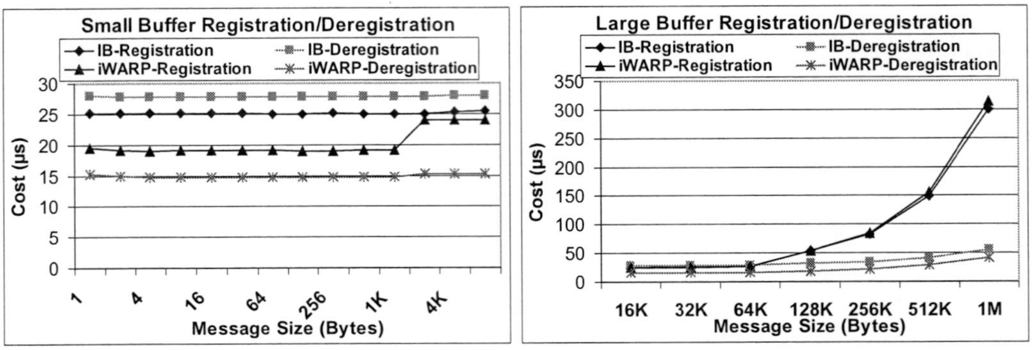

Figure 10. Memory registration/deregistration cost.

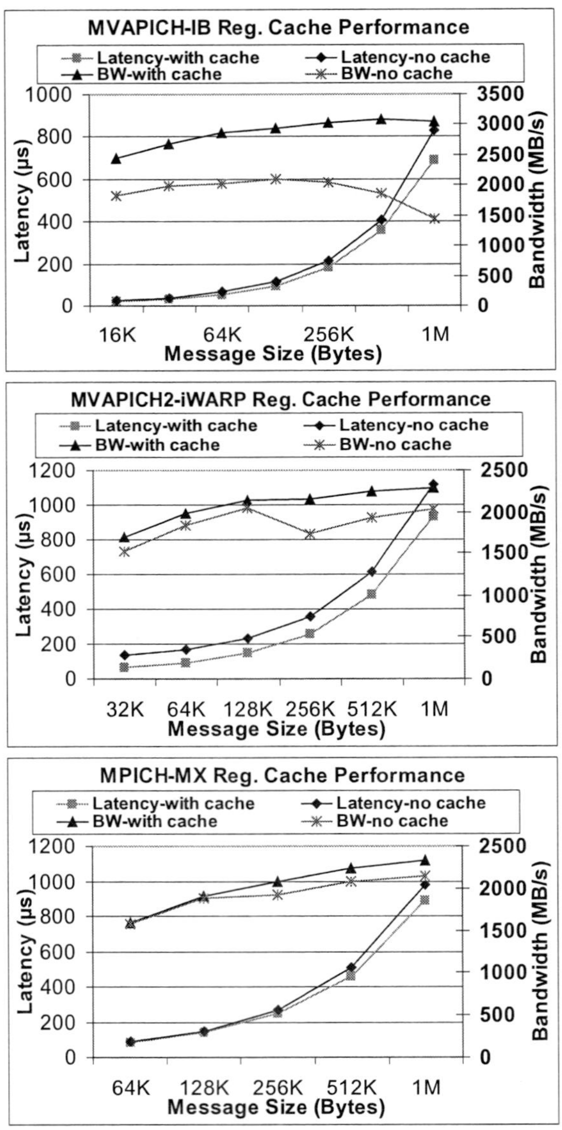

Figure 11. Performance effect of Memory registration cache on MPI latency and bandwidth.

To avoid re-registration costs at the verbs layer in future buffer reuses, MVAPICH-IB and MVAPICH2-iWARP implementations use a lazy de-registration mechanism [28]. In this mechanism, each previously registered buffer remains registered and its buffer address is stored in a registration cache. Instead of actually registering and deregistering the buffer, its cache reference count is increased/decreased every time the buffer is needed to be registered/deregistered. Only when the registration cache is full, a buffer with zero reference count will be evicted from the cache and then physically deregistered. To keep the information in the registration cache coherent with the operating system virtual memory changes, a synchronization with the operating system kernel is required that may negatively affect the performance of the registration cache. The MX-10G library uses a similar internal mechanism at the user-level. In case that an application frequently reuses its buffers, the use of registration cache can help avoid repeating the registration of a buffer unnecessarily.

It is therefore worth measuring the actual performance of memory registration cache used in the MPI implementations for these interconnects. This way we can determine how useful the registration cache can be for interconnects that require explicit or implicit memory registrations. Figure 11 shows the impact of disabling/enabling the registration cache on the latency and bandwidth of MPI implementations over InfiniBand, iWARP and Myrinet networks, when we reuse all the buffers used for communication in the benchmark. We report the ping-pong results for the latency and the both-way results for the bandwidth. The message sizes used for measurement are all in the Rendezvous protocol range, where buffer registration is performed. As shown in Figure 11, the effect of using registration cache on latency and bandwidth is considerable. The effect on Myrinet network, for both latency and bandwidth tests is relatively lower than the other two interconnects, which shows a relatively low buffer registration cost for this network. Since memory registration and the registration cache are internal to the MX library, this feature is a characteristic of the MX library and the Myrinet card.

7.2. Effect of Message Buffer Reuse

Reusing a previously referenced buffer has some important performance implications in modern computing and communication architectures. Applications using different message buffers may need to register/deregister their user buffers. Application's buffer reuse pattern affects the registration cache performance, leading to a significant impact on the overall communication performance. Clearly, memory/cache performance, memory address translation overhead and TLB performance are also affected with the application buffer reuse pattern.

In the previous section, we evaluated the effect of memory registration and registration cache on the performance of a micro-benchmark that completely reuses its buffers, and we observed that the effect of using the registration cache is significant. In this section, we want to show how reusing communication buffers can affect the performance in the existence of registration cache.

To evaluate the impact of buffer reuse on communication performance, we examine different buffer reuse patterns for the MPI ping-pong test. In this test, we statically allocate a number of separate memory buffers and depending on the buffer reuse pattern we select a new buffer from the available buffers or use the previously used buffer. Figure 12 shows the ping-pong latency when changing the buffer reuse pattern from no-reuse (0% buffer reuse) to half-reuse (50% buffer reuse) and full-reuse (100% buffer reuse or always use a single buffer) for each network. Figure 12 also shows the ratio of no-reuse latency to full-reuse latency.

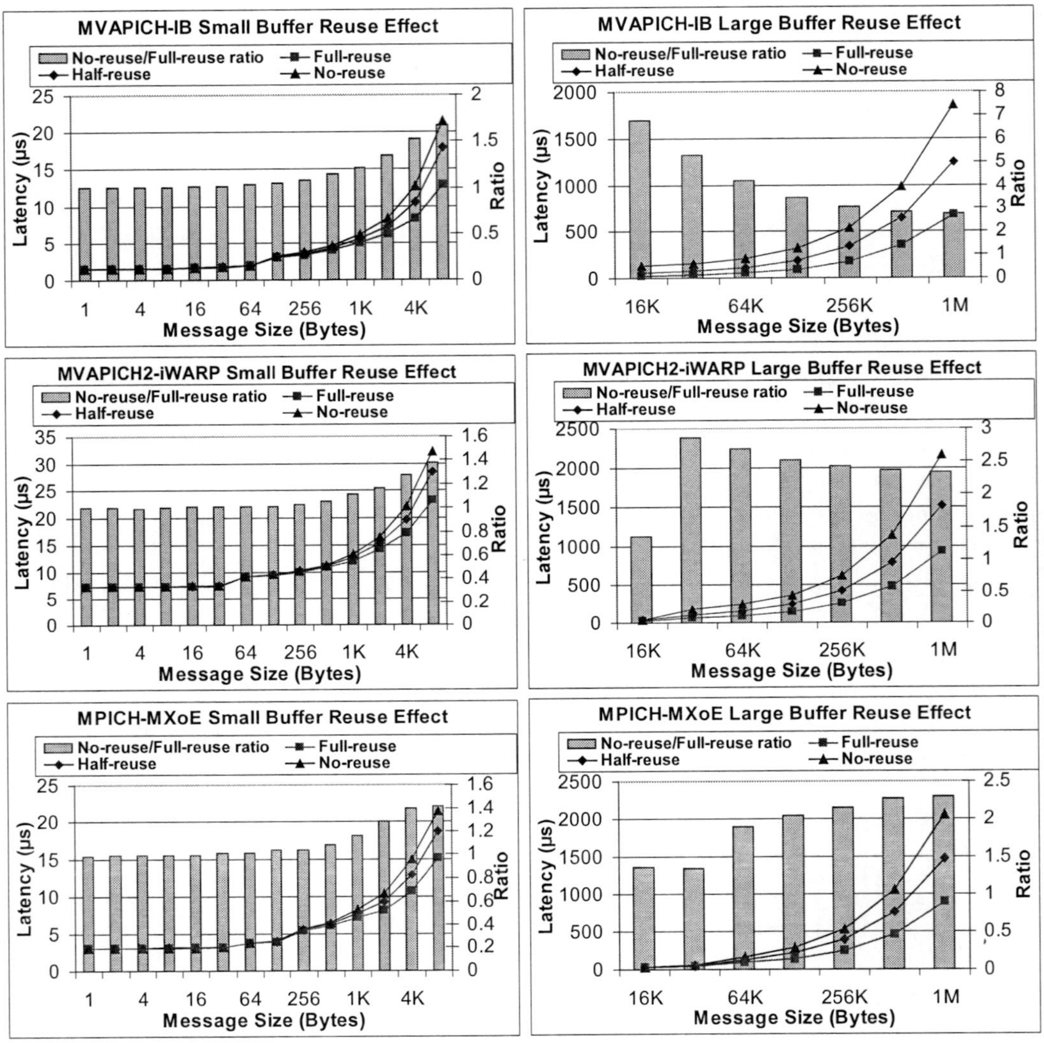

Figure 12. Effect of buffer re-use.

For small messages up to 512 bytes, we see less than 20% impact for all the networks. For Eager size messages, this ratio is less than 1.4 for iWARP, 1.7 for IB and 1.9 for Myrinet. The ratio grows for Rendezvous size messages and reaches 6.7 for IB at 16KB, 2.8 for iWARP at 32KB, and around 2.3 for Myri-10G network at 1MB. This will have a large effect on the Rendezvous protocol's performance, especially when changing the communication buffers more frequently. All these networks require costly memory registrations for the Rendezvous protocol at the sender and receiver sides when new message buffers are used. However, the impact on the IB is more significant, which can be attributed to its lower base latency compared to the other interconnects.

For large messages, no data copy is performed in the Rendezvous protocol. However, both the sender and receiver processes first search the pin-down cache for a pre-registered buffer. If the buffer is not found in the cache, it needs to be registered. Therefore, with a low buffer reuse profile, a considerable impact due to memory registration cost is observed. We should also add the effect of TLB and cache flushing due to using different user buffers. As observed in Figure 12, for all interconnects there is a jump in the ratio of no-reuse latency to full-reuse latency at their Eager/Rendezvous switching point.

On the other hand, for the Eager protocol, source buffers are not directly used for data transfer. Instead, messages are first copied into a previously registered staging buffer. Similarly, messages are first received into pre-registered buffers, and then copied into their destination buffers. This method avoids the registration cost at the expense of two extra copies. So the effect on Eager messages is lower compared to the Rendezvous messages. Nevertheless, there is a slight effect, increasing with the buffer size, which is due to cache and TLB effects.

8. Effect of MPI Queue Usage

MPI implementations use several queues for processing communication calls. The receive call queue is used to keep early posted receive calls. It is traversed upon reception of a message from the network to find a matching receive call. The unexpected message queue is used to save unexpected messages temporarily (messages that no matching receive call has yet been posted for). When a receive call is invoked, the unexpected message queue is traversed for a matching message. If no unexpected message is found, a new receive request will be posted to the receive call queue. In this section, we examine the effect of queue usage for these two queues. For InfiniBand and iWARP networks, these two queues are manipulated inside the MPI code. Myrinet manages these queues in its MX user-level library.

8.1. Unexpected Message Queue

To measure the effect of unexpected queue usage, we use an algorithm similar to [49]. Each process sends a certain number of small, unexpected messages to the other side. Then the processes synchronize and start communicating in a ping-pong fashion. We modified the algorithm in [49] and used blocking send calls instead of non-blocking calls to minimize overlapping of queue processing with message communication time, in order to measure the worst-case latency effects.

In Figure 13, we show the message latency in three queue sizes: full queue (2000 unexpected messages), half queue (1000 unexpected message) and empty queue. Figure 13 also shows the ratio of full queue latency to empty queue latency. As shown, small and medium size messages are considerably affected when 2000 unexpected messages exist in the queue. This is because the queue search time is large compared to the message latency. Since in the Rendezvous range the message latency is higher than searching the unexpected queue for the matching message (which is in the form of a control message rather than the actual data), the effect on large messages is insignificant.

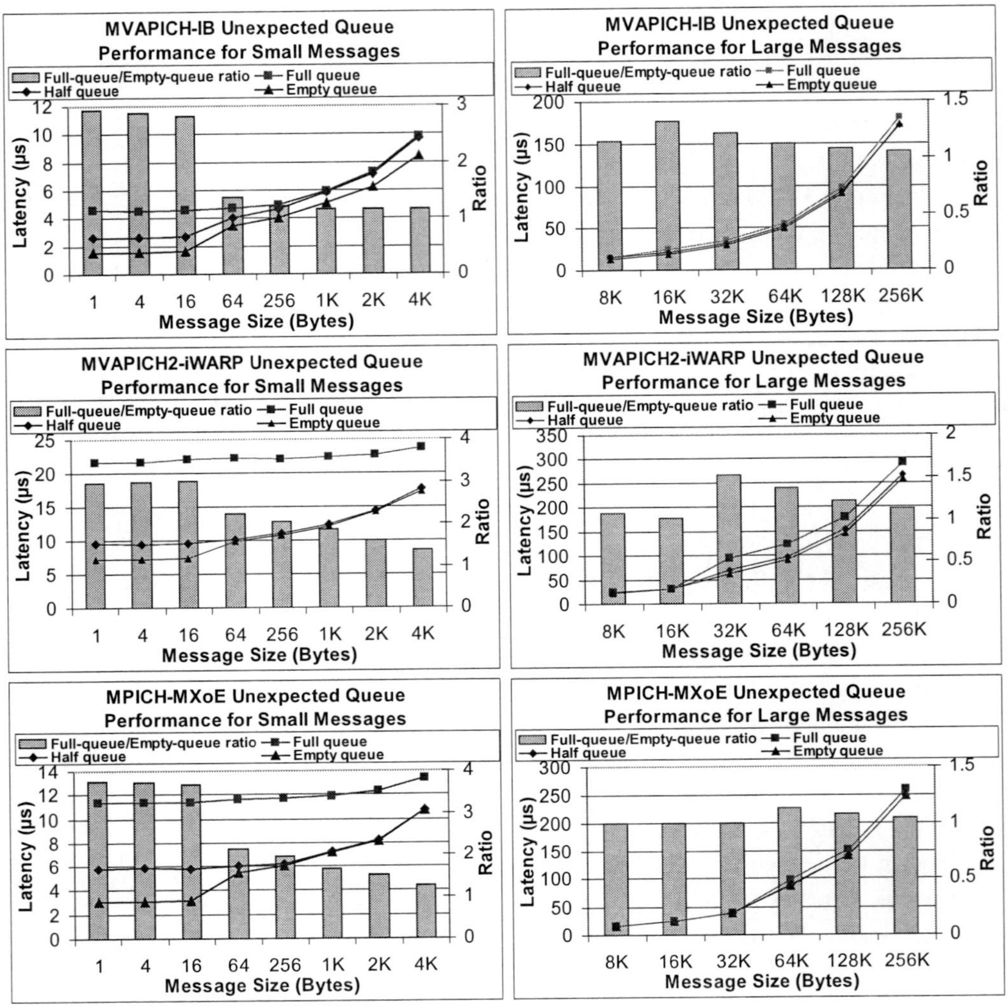

Figure 13. Effect of unexpected messages.

8.2. Receive Call Queue

For the receive call queue, we use an algorithm similar to the one in [49]. As shown in Figure 14, initially, both sides of the communication pre-post a certain number of non-blocking receive calls with a certain tag (tag1), which are called "traversed calls". These calls sit at the beginning of the receive call queue. Then the actual latency-measured non-blocking receive call is posted with a different tag (tag2). This call sits in the queue after the traversed calls. Some other receive calls with tag3 are posted afterwards ("non-traversed calls"). At this stage, both sides synchronize and start communicating. One side sends a message with tag2 and waits for a similar message in response. Upon reception of a message at either side, the queue is traversed to find the matching receive with tag2. Before finding the matching call, all pre-posted calls with tag1 are traversed (but not processed).

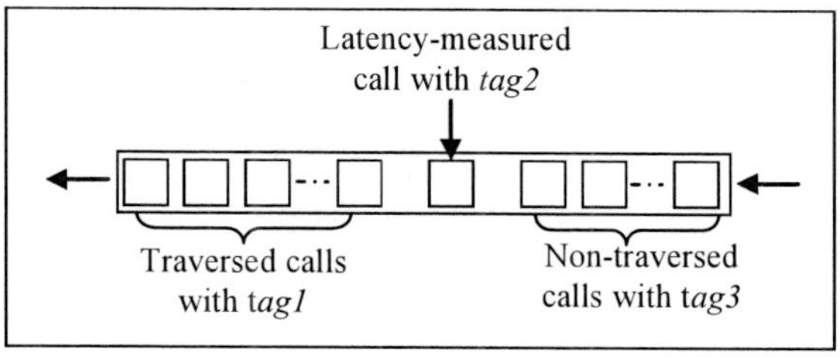

Figure 14. Receive call queue test.

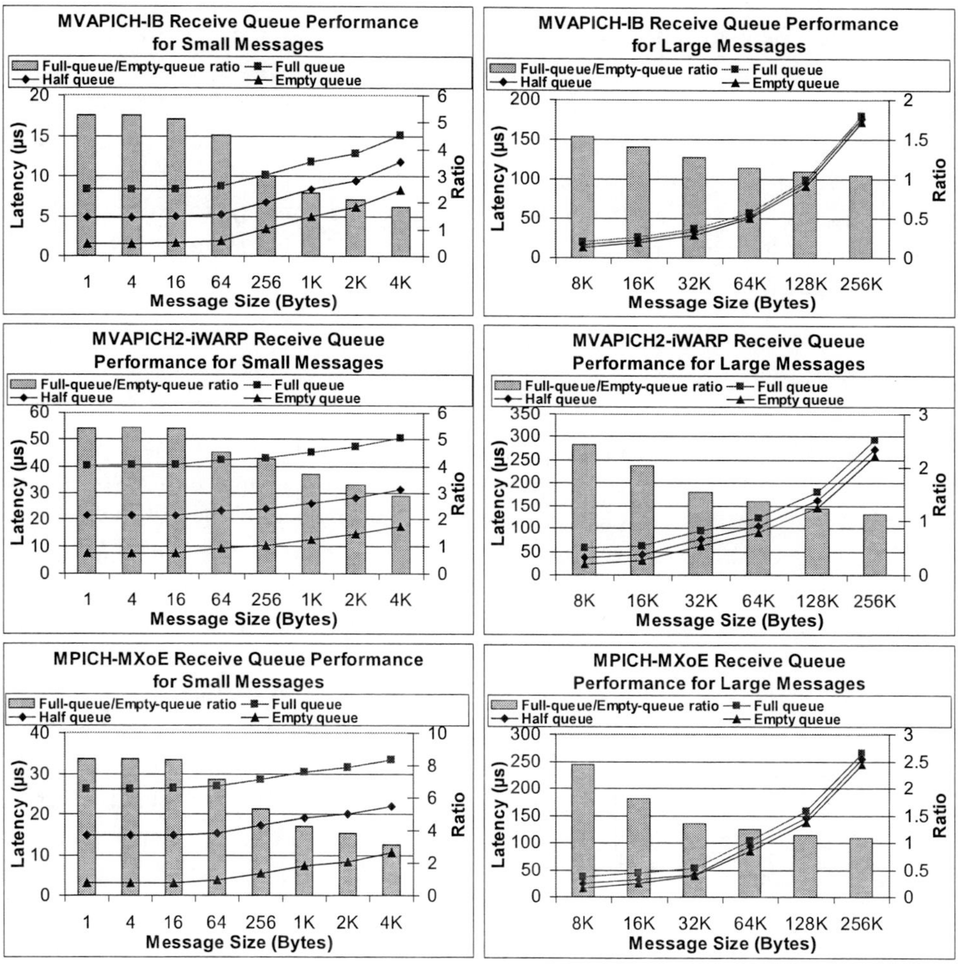

Figure 15. Effect of receive queue usage.

In Figure 15, we show the latency when there are 2000 traversed calls (full queue), 1000 traversed calls (half queue) and no traversed calls (empty queue). Figure 15 also shows the ratio of full queue latency to empty queue latency. The small (Eager size) message latency is affected more with the increasing size of pre-posted receive queue, which is because queue search time becomes larger than the communication latency itself. Obviously the queue search time effect diminishes with increasing message size, until for very large messages the message latency in the full queue case becomes equal to the latency in the empty queue case.

9. Overlap and Progress

Overlapping computation with communication is one of the basic techniques in hiding communication latency, thereby improving application performance. Using non-blocking communication calls at the application level, supporting independent progress for non-blocking operations at the messaging layer, and offloading communication processing to the NICs are the main steps in achieving efficient communication/computation overlap.

NICs in modern interconnects are designed to offload most of the network processing tasks from the host CPU, providing excellent opportunity for communication libraries such as MPI to hide the communication latency using non-blocking calls. To utilize the offload engines efficiently, non-blocking communications need to make progress independently. While some MPI implementations support independent progress, others require subsequent library calls in order to make progress in outstanding non-blocking calls. This may have a significant impact on performance when a computation phase follows a non-blocking call. In this section, we intend to assess the ability of modern interconnects to independently make progress in outstanding non-blocking communications and consequently their ability to overlap communication with computation.

9.1. Computation/Communication Overlap

We conduct our overlap measurements using micro-benchmarks for both MPI non-blocking send and receive operations. We use separate benchmarks to measure send and receive overlap ability. This way we can clearly identify the reasons behind the observed behavior at both sender and receiver sides. In our tests, unlike the method used in [12] we directly measure the amount of computation that can be overlapped with communication. Since the MPI libraries under study use Eager and Rendezvous protocols to transfer small and large messages respectively, we analyze the results for small and large messages separately.

In the send overlap micro-benchmark, the sender uses a loop starting with a non-blocking send call and followed by a computation phase. It then waits for the completion of the message transfer using MPI_Wait(). The receiver blocks on an MPI_Recv(). The time is measured at the sender side. In the receive overlap micro-benchmark, the receiver posts a non-blocking receive call, MPI_Irecv(), and then computes. It then waits for the communication to complete using MPI_Wait(). The sender blocks on an MPI_Send(). Time is measured at the receiver side.

To calculate the overlap in each iteration of the loop, we measure the portion of the computation phase that can be fully overlapped with communication, without affecting the communication time. Let l_0 be the communication time with no inserted computation, c_m, the computation time inserted in the iteration m, and l_m, the communication time of the iteration m. We increase the amount of computation, c_m, by 10% in each iteration until 10% increase in the original communication time, l_0, is observed (the 10% values are approximate numbers, selected based on empirical observations). The last iteration is the largest m, with $l_m < (1.1)\ l_0$.

Figure 16 illustrates the timing measurement of our overlap micro-benchmark. In this model, t_1 is the period that the non-blocking send/receive call is active. In fact, the host CPU is busy with communication processing during t_1. Obviously, the best-case scenario is when the non-blocking call is able to start the communication in the NIC offload engine, before returning from the call. The t_2 period starts when the non-blocking call returns. In this period, the NIC is performing the data transfer using its offload engine, and the CPU is available for computation. Clearly, t_2 is the upper bound for overlap. The t_3 period is the time that the progress engine, called by MPI_Wait(), will complete the communication. It starts right after the computation phase (c_m) or the offload phase (t_2), whichever finishes later.

The original communication latency can be calculated using Equation (1). Clearly, if the computation phase in iteration m (c_m) is smaller than t_2, it will be fully overlapped with the communication, and therefore l_m would be equal to l_0, as in Equation (2). It is clear that the overlapped computation amount is equal to c_m. On the other hand, if c_m is greater than t_2 (not shown in Figure 16), since no more than t_2 time is available for overlap, the start of t_3 period will be delayed, consequently increasing the l_m, as in Equation (3). Clearly, in this case, the overlap time is equal to t_2.

$$l_0 = t_1 + t_2 + t_3 \tag{1}$$

$$l_m = t_1 + t_2 + t_3 = l_0 \tag{2}$$

$$l_m = t_1 + c_m + t_3 \tag{3}$$

Using the above equations for iteration m, one can derive the communication/computation overlap time for either of the discussed cases, as c_m - $(l_m$- $l_0)$. Therefore, the overlap ratio is as shown in equation 4:

$$overlap\_ratio = \frac{c_m - (l_m - l_0)}{l_0} \tag{4}$$

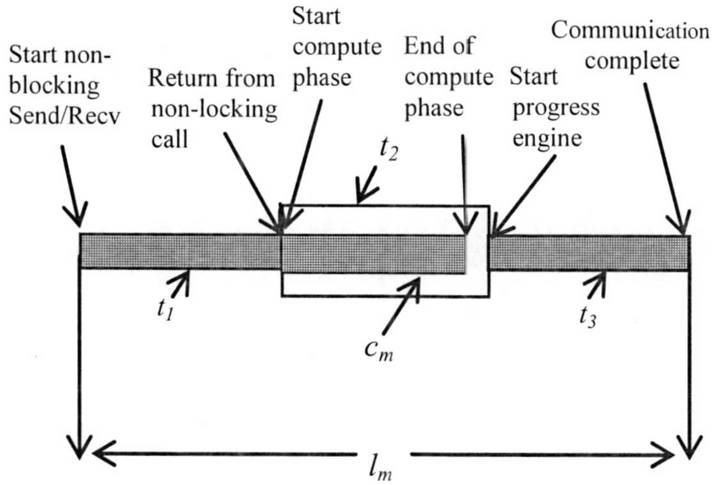

Figure 16. Timing model for overlap benchmark.

9.1.1. Send Side Overlap

Figure 17 illustrates the send and receive communication/computation overlap ability of the networks with their MPI implementations. For InfiniBand and iWARP, we have evaluated two Rendezvous protocols, RDMA Read based (RGET) and RDMA Write based (RPUT) [48].

One can observe a different behavior for small (Eager range) and large (Rendezvous range) messages. For sending Eager size messages, all networks show a high level of overlap ability. For the Rendezvous protocol, MPICH-MXoE, MVAPICH-IB and MVAPICH2-iWARP using RGET protocol maintain their high overlap ability, and even reach higher overlap percentages. In the RGET-based Rendezvous protocol over iWARP and IB, the overlap ability increases for Rendezvous-size messages and approaches 100% for 1MB messages. This is because in the RGET Rendezvous protocol, after a one-way handshake from the sender called *Request to Send* (RTS), the receiver uses RDMA Read to retrieve the message from the sender's memory and therefore the sender is almost free during this operation. Similarly, for Myrinet network, data are transferred by a direct Get initiated by the receiver using a progression thread, which makes both the sender and receiver processors available [31]. This makes Myrinet send overlap very high for large messages.

On the other hand, both the MVAPICH-IB and MVAPICH2-iWARP results show significant degradation where the send overlap becomes very low for large messages, when using RPUT Rendezvous protocol. This is primarily because RPUT uses a two-way handshake protocol followed by RDMA Write for data transfer [48]. In this protocol, the sender sends an RTS to the receiver. The receiver will reply with a *Clear To Send* (CTS) message that enables the sender to start the data transfer. Therefore, when the sender enters a computation phase after its non-blocking call, the negotiation cycle (reception of CTS) remains incomplete, delaying the start of data transfer until the next MPI communication call at the sender side.

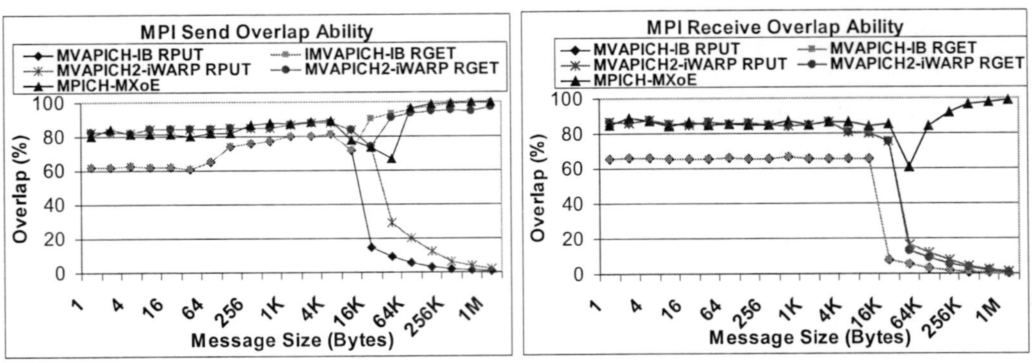

Figure 17. Communication/computation overlap.

9.1.2. Receive Side Overlap

In all networks, as shown in Figure 17, a high level of overlap ability exists for the MPI receive operation up to Eager/Rendezvous switching point. For InfiniBand and iWARP, after the switching point, this overlap ability drops to a very low percentage, regardless of the Rendezvous protocol type. Obviously, for large messages, the non-blocking receive calls do not start the data transfer, serializing computation and communication phases. This is due to inefficient use of MPI progress engine as well as the nature of the used Rendezvous protocol. In case of an early-arrived receiver, no RTS is arrived and thus the progress remains after returning from the computation phase. In case that the receiver arrives later than the sender, a progress engine call is needed in the non-blocking receive call to recognize the RDMA-transferred RTS message. Since in none of the Rendezvous protocol implementations such a call exists, the data transfer for two different protocols remains after the computation phase, decreasing the overlap.

A different behavior is observed though for large messages over Myrinet. Since most of the data transfer for Rendezvous-rage messages is started with a progression thread [31], the receiver CPU is free during the data movement and thus the communication can asynchronously make progress while the CPU is in computation phase.

9.2. Communication Progress

Communication progress is called independent if a pending non-blocking communication can make progress without subsequent library calls. This ability helps overlap the on-going communication with computation. The level of independent progress depends on the communication protocols, the progress engine and the underlying hardware offload support. Generally, the progress can be made in protocol negotiations and/or data transfer. Inevitably, with the existence of offload, data movement can proceed independently from the host CPU. However, the start of data transfer and the completion of communication protocol may still need further library calls.

While most MPI implementations use polling-based progress engines, some have the option of using interrupts for communication progress [48][23]. In polling-based progress, even in the existence of offload, any computation phase that follows a non-blocking call may delay the progression of the communication protocol that has already been started by the non-blocking call. Although interrupt-based approaches activate the progress engine any time communication progress is needed, which leads to independent progress while the computation is performed in the CPU, they impose some overhead and fluctuation on the communication time, which cannot be overlooked.

9.2.1. Sender-Side Progress

In our send progress micro-benchmark, the sender invokes a non-blocking send call after synchronization with the receiver. It then spends a certain amount of time in a synthetic delay routine. Unlike the tests for measuring overlap, the inserted computational delay is chosen to be longer than the basic message latency to clearly highlight the progress ability. After the delay, MPI_Wait() is called for the completion of the send request. Then the sender waits for acknowledgement from the receiver. At the other end and after synchronization, the receiver calls a blocking receive operation and then acknowledges the reception by sending a small message back to the sender. The overall receive latency is recorded for several inserted delay values for each message size. If the latency is increased by the inserted delay, then the communication has not made progress independently.

Figure 18 depicts the measured results for the send communication progress in MPI implementations over four interconnects. For InfiniBand and Myrinet, the latency for Eager size messages (8KB in this test) is not affected by the inserted delay. For iWARP the effect is small, and in fact the message latency is not shifted by the huge amount of synthetic delay. Results show that the communication completes during the delay time without any subsequent MPI calls at the sender side.

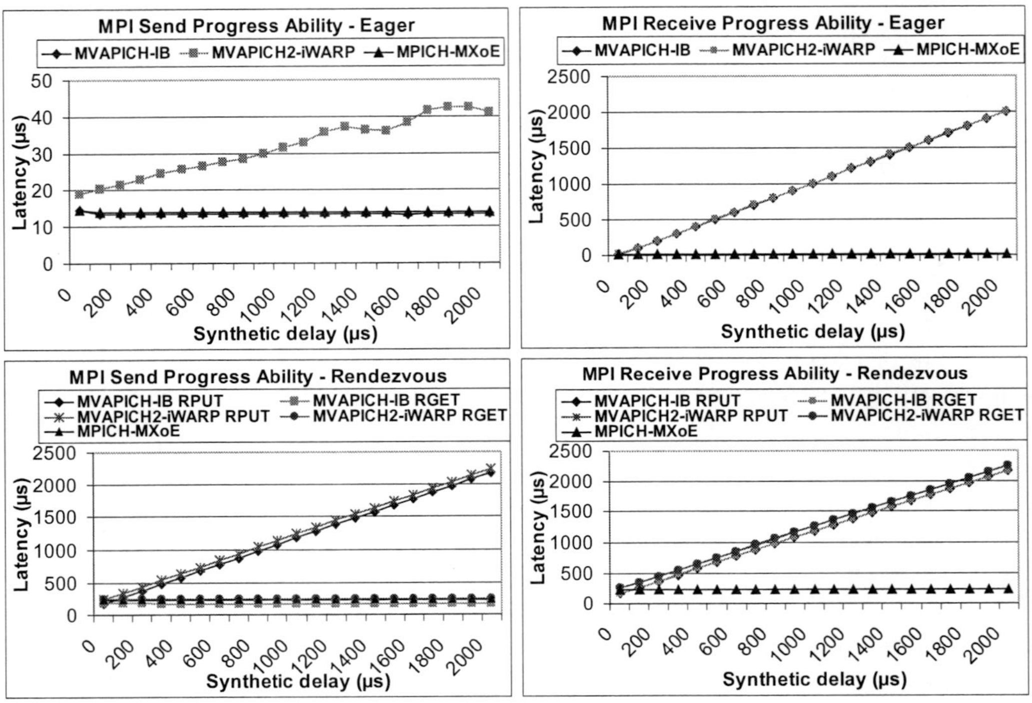

Figure 18. MPI communication progress results.

For the Rendezvous range (256KB in this test), Myrinet and both InfiniBand and iWARP when using RGET protocol show a flat latency curve, meaning that the communication is independently making progress during the delay period. Here, in a one-way Rendezvous negotiation, the receiver retrieves the data from the sender using RDMA Read upon reception of RTS from the sender [48]. Therefore, the receiver can start the data transfer, independently making progress by utilizing offloaded protocol engine. That is why we see a flat delay curve for large messages. On the other hand, when using RPUT protocol for IB and iWARP, the latency is affected by the inserted delay making the completion of communication happen after the delay. Here, even the completion of the two-step Rendezvous negotiation remains after the delay, thus no data are being transferred during the inserted synthetic delay period.

Comparing the send-side overlap and progress results shows a correlation between overlap and progress ability of a network in each case. All networks show a relatively high overlap and progress ability when sending small messages. RGET-based Rendezvous protocols show high overlap and progress ability, while RPUT-based Rendezvous protocols over IB and iWARP are relatively weak in communication progress and show low overlap abilities.

9.2.2. Receiver-Side Progress

We run a similar test for the receive operation. The timing measurement is again performed at the receiver side. After a barrier synchronization, the receiver posts a non-blocking receive call and enters a synthetic delay. During the delay, it also checks for probable completion of the message reception by examining the receive buffer for a certain data value. This helps detect whether the message has been completely received while the receiver is in the delay period. After the delay, MPI_Wait() is called to complete the request. At the send side, a blocking send is called after the synchronization.

Results of the receive progress micro-benchmark for MPI libraries under study are also shown in Figure 18. Clearly, except for Myrinet, none of the MPI implementations shows independent progress for receiving Eager size messages (8KB in this test), and their completion is delayed by the inserted delay time. In general, Eager size messages in MPICH-MXoE are transferred to a temporary buffer at the receive side upon posting the MPI send call. Similar to what we discussed in Section 9.1.2 about progress engine call, the observed delay effect is because the reception of the message is not checked in the non-blocking receive call and is left to the progress engine in future MPI calls. Thus, the completion of the receive operation remains for the time that we call MPI_Wait(), which is in fact after the inserted delay.

Examining the details of receive latencies for Eager-size messages in all networks confirms that only a small portion of the original (non-delayed) communication latency remains after the inserted delay. For instance, the amount of post-delay work for iWARP is only 25% of the original message latency for 8KB messages. This implies that the actual data transfer has been performed during the delay (using direct memory transfer), but the post processing has remained for the MPI progress engine, which is invoked after the delay. This post processing includes finding the arrived message in the unexpected queue, copying it into the user buffer and finalizing the receive data structure. The reason that we do not see this effect for Myrinet network is that in case a matching receive is provided when the data arrives at the receiver, the data are directly put into the receive buffer by the MX library. This is unlike RDMA-based transfers in MPI implementations over IB and iWARP for which the arrived data needs to be copied out of a temporary buffer.

The case for the Rendezvous protocol is quite different. For IB and iWARP, the results look similar to that of Eager case, but the latency details imply a different story. In fact, regardless of the Rendezvous protocol in use (RGET or RPUT), all of the message latency remains after the inserted delay. This means that, unlike the discussion in Section 9.2.1 for send progress, even the data transfer does not start until the inserted delay ends. This is primarily because the non-blocking (send or receive) call does not find the Rendezvous control message (CTS or RTS) to start the data transfer. These messages will be recognized only at the next progress engine execution. The case for Myrinet is different though. Due to using asynchronous progress using a progression thread, a very small processor overhead on CPUs at both sides is imposed [31], making the CPUs almost free for computation. The receive side progress results are also in concert with the corresponding receive overlap results explained in Section 9.1.2. Wherever we have high overlap ability, high level of independent progress at least in data transfer exists.

10. Conclusion

In this work, we have shown the benefits of using high-performance interconnects with their user-level messaging libraries over regular Ethernet communication stack. Results show that very low latencies and very high bandwidth results are achievable using modern interconnects. For example, around eight times less latency and up to three times more bandwidth are achieved using modern communication stacks over high-performance interconnects. Modern networks are also showing different but relatively good scalability in latency and bandwidth when multiple active connections exist.

This chapter has also analyzed the various features of MPI implementations over modern HPC interconnects. Due to low network latency, MPI implementations of these modern interconnects are very sensitive to host based processing overheads such as low buffer reuse overheads or lengthy MPI queues. For example, the small message latency can be increased several times if a high number of entries are queued in the MPI unexpected or posted receive queues. In addition to low network latency, high memory registration cost makes the MPI implementations sensitive to buffer reuse pattern and registration cache performance. As our results confirm, unlike MPI queue processing effect, the buffer reuse and memory registration effect is higher for the Rendezvous protocol range (large messages). Therefore reusing large message buffers in applications plays a significant role in their communication performance on RDMA-based interconnects.

In this chapter, we also assessed the ability of the modern cluster interconnects with offload capability and their MPI libraries for communication/computation overlap and communication progress. In terms of overlap, all of the MPI libraries are able to overlap computation with sending small messages that are transferred eagerly. All libraries also show a high level of overlap ability for sending large messages, where a short (one-step) negotiation is used.

The MPI communication progress results also confirm that without independent progress (at least in data transfer) we cannot achieve high level of overlap. MVAPICH2-iWARP and MVAPICH-IB that have shown a good send overlap ability show independent progress for send operation as well. Myrinet also shows a good progress for sending messages, and is able to make independent progress due to its asynchronous progress ability.

All networks but Myrinet show some level of receive progress for small messages, although it is just in data transfer. MPI over iWARP and InfiniBand show poor receive progress for large messages because the Rendezvous negotiation (reception and recognition of RTS at the receiver side) does not complete in non-blocking calls, preventing the data transfer to start. Myrinet has a better receive progress for all message sizes.

Acknowledgements

This research is supported by the Natural Sciences and Engineering Research Council of Canada through grant RGPIN/238964-2005, Canada Foundation for Innovation's grant #7154, and Ontario Innovation Trust's grant #7154. The Authors would like to thank NetEffect Inc. and Mellanox Technologies Inc. for the resources. We acknowledge Myricom for providing technical support.

References

[1] Argonne National Laboratory. MPICH2. Available: http://www-unix.mcs.anl.gov/mpi/mpich2/

[2] Argonne National Laboratory. MPICH. Available: http://www-unix.mcs.anl.gov/mpi/mpich1

[3] J. Beecroft, D. Addison, D. Hewson, M. McLaren, D. Roweth, F. Petrini and J. Nieplocha, "QsNetII: Defining High-Performance Network Design," *IEEE Micro*, vol. 25, pp. 34-47, 2005. doi:10.1109/MM.2005.75

[4] C. Bell, D. Bonachea, Y. Cote, J. Duell, P. Hargrove, P. Husbands, C. Iancu, M. Welcome and K. Yelick, "An Evaluation of Current High-Performance Networks," *Proceedings of the 17th IEEE International Parallel and Distributed Processing Symposium (IPDPS),* 2003. doi:10.1109/IPDPS.2003.1213106

[5] R. Brightwell, D. Doerfler and K. D. Underwood, "A comparison of 4X InfiniBand and quadrics elan-4 technologies," in *Proceedings of the IEEE International Conference on Cluster Computing (IEEE Cluster),* 2004, pp. 193-204. doi:10.1109/ CLUSTR.2004.1392617

[6] R. Brightwell, R. Riesen and K. D. Underwood, "Analyzing the Impact of Overlap, Offload, and Independent Progress for Message Passing Interface Applications," *International Journal of High Performance Computing Applications*, vol. 19, pp. 103-117, 2005. doi:10.1177/ 1094342005054257

[7] ConnectX InfiniBand Adapters, product brief, Mellanox Technologies, Inc., Available: http://www.mellanox.com/pdf/products/hca/ConnectX_IB_Card.pdf

[8] P. Culley, U. Elzur, R. Recio and S. Bailey. Marker PDU aligned framing for TCP specification (version 1.0). RDMA Consortium, Available: http://www.rdmaconsortium.org/home/draft-culley-iwarp-mpa-v1.0.pdf

[9] D. Dalessandro and P. Wyckoff, "A performance analysis of the Ammasso RDMA enabled Ethernet adapter and its iWARP API," in *Proceedings of the RAIT Workshop at IEEE International Conference on Cluster Computing (IEEE Cluster),* 2005, doi:10.1109/CLUSTR.2005.347028

[10] D. Dalessandro, P. Wyckoff and G. Montry, "Initial performance evaluation of the NetEffect 10 gigabit iWARP adapter," in *Proceedings of the RAIT Workshop at the IEEE International Conference on Cluster Computing (IEEE Cluster),* 2006, doi:10.1109/CLUSTR.2006.311915

[11] DAT Collaborative. Available: http://www.datcollaborative.org/

[12] D. Doerfler and R. Brightwell, "Measuring MPI send and receive overhead and application availability in high performance network interfaces," in *Proceedings of the 13th Euro PVM/MPI Conference*, 2006, pp. 331-338. doi:10.1007/11846802

[13] W. Feng, P. Balaji, C. Baron, L. N. Bhuyan and D. K. Panda, "Performance characterization of a 10-Gigabit Ethernet TOE," in *Proceedings of the 13th Symposium on High Performance Interconnects (HOTI '05),* 2005, pp. 58-63. doi:10.1109/CONECT.2005.30

[14] J. Hilland, P. Culley, J. Pinkerton and R. Recio. RDMA protocol verbs specification (version 1.0). RDMA Consortium, Available: http://www.rdmaconsortium.org/home/draft-hilland-iwarp-verbs-v1.0-RDMAC.pdf

[15] J. Hurwitz and W. Feng, "Analyzing MPI performance over 10-Gigabit Ethernet," *Journal of Parallel and Distributed Computing*, vol. 65, pp. 1253-1260, 2005. doi:10.1016/j.jpdc.2005.04.011
[16] *IETF: The Internet Engineering Task Force*. Available: http://www.ietf.org/
[17] Internet Engineering Task Force (IETF), "Stream Control Transmission Protocol (SCTP)," RFC 2960, Available: http://www.ietf.org/rfc/rfc2960.txt
[18] InfiniBand Trade Association. InfiniBand architecture specification. Available: http://www.infinibandta.org/
[19] W. Jiang, J. Liu, H. Jin, D. K. Panda, W. Gropp and R. Thakur, "High performance MPI-2 one-sided communication over InfiniBand," in *IEEE/ACM International Symposium on Cluster Computing and the Grid (CCGrid),* 2004, pp. 531-538. doi:10.1109/CCGrid.2004.133664
[20] T. Kielmann, H. E. Bal and K. Verstoep, "Fast measurement of LogP parameters for message passing platforms," in *Proceedings of the 15th IEEE International Parallel and Distributed Processing Symposium (IPDPS),* 2000, pp. 1176-1183. doi:10.1007/3-540-45591-4
[21] H.-. Jin, S. Narravula, G. Brown, K. Vaidyanathan, P. Balaji and D. K. Panda, "Performance evaluation of RDMA over IP: A case study with the Ammasso gigabit Ethernet NIC," in *Workshop on High Performance Interconnects for Distributed Computing at the 14th International Symposium on High Performance Distributed Computing*, 2005.
[22] M. Koop, J. Sridhar and D.K. Panda, "Scalable MPI Design over InfiniBand using eXtended Reliable Connection," *Proceedings of the IEEE Conference on Cluster Computing (IEEE Cluster)*, 2008. doi:10.1109/CLUSTR.2008.4663773
[23] R. Kumar, A.R. Mamidala, M.J. Koop, G. Santhanaraman and D.K. Panda. Lock-free asynchronous Rendezvous design for MPI point-to-point communication. In *Proceedings of the 15th Euro PVM/MPI Conference*, LNCS 5205, 2008, pp. 185-193. doi:10.1007/978-3-540-87475-1_27
[24] J. Liu, B. Chandrasekaran, J. Wu, W. Jiang, S. Kini, W. Yu, D. Buntinas, P. Wyckoff and D. K. Panda, "Performance comparison of MPI implementations over InfiniBand, Myrinet and quadrics," in SC '03: *Proceedings of the 2003 ACM/IEEE Conference on Supercomputing,* 2003, pp. 58-71. doi:10.1109/SC. 2003.10007
[25] J. Liu, B. Chandrasekaran, W. Yu, J. Wu, D. Buntinas, S. Kini, D. K. Panda and P. Wyckoff, "Microbenchmark Performance Comparison of High-Speed Cluster Interconnects," *IEEE Micro,* vol. 24, pp. 42-51, 2004. doi:10.1109/MM.2004.1268994
[26] Mellanox Technologies. Available: http://www.mellanox.com
[27] F. Mietke, R. Rex, R. Baumgartl, T. Mehlan, T. Hoefler and W. Rehm, "Analysis of the memory registration process in the Mellanox InfiniBand software stack," in *Proceedings of 12th International Euro-Par Conference*, Dresden, Germany, 2006, pp. 124-133. doi:10.1007/11823285_13.
[28] F. Mietke, R. Rex, T. Mehlan, T. Hoefler, and W. Rehm, "Reducing the impact of memory registration in InfiniBand", in the 1st Kommunikation in Clusterrechnern und Clusterverbundsystemen (KiCC), 2005.
[29] MPI Forum, "MPI: A message-passing interface standard," MPI Forum, 2003.
[30] MPI over InfiniBand Project. Available: http://mvapich.cse.ohio-state.edu/
[31] Myricom Homepage. Available: http://www.myricom.com

[32] S. Narravula, A. R. Mamidala, A. Vishnu, G. Santhanaraman and D. K. Panda, "High performance MPI over iWARP: Early experiences," in *Proceedings of the IEEE International Conference on Parallel Processing (ICPP),* 2007, pp. 46-53. doi:10.1109/ICPP.2007.46

[33] NetEffect Inc. NetEffect NE010 10-Gigabit iWARP Ethernet channel adapter data sheet. Available: http://www.neteffect.com/documents/NE010_Product_Brief_060806.pdf

[34] NetEffect Inc. NetEffect NE020 10-Gigabit iWARP Ethernet channel adapter. NetEffect Inc., Available: http://www.neteffect.com/documents/NE020_Product_Brief.pdf

[35] OpenFabrics Alliance Homepage. Available: http://www.openfabrics.org/

[36] F. Petrini, E. Frachtenberg, A. Hoisie and S. Coll, "Performance Evaluation of the Quadrics Interconnection Network," *Cluster Computing*, vol. 6, pp. 125-142, 2003. doi:10.1023/A:1022852505633

[37] Y. Qian and A. Afsahi, "Efficient Shared Memory and RDMA based Collectives on Multi-rail QsNetII SMP Clusters", Cluster Computing, *The Journal of Networks, Software Tools and Applications,* vol. 11, No. 4, December 2008, pp 341-354. doi:10.1007/s10586-008-0065-8

[38] Y. Qian, A. Afsahi and R. Zamani, "Myrinet networks: A performance study," in Proceedings of the Network Computing and Applications, *Third IEEE International Symposium on (NCA'04),* 2004, pp. 323-328. doi:10.1109/NCA.2004.1347794

[39] M. J. Rashti and A. Afsahi, "10-Gigabit iWARP Ethernet: Comparative performance analysis with InfiniBand and Myrinet-10G", in the *21st IEEE International Parallel and Distributed Processing Symposium (IPDPS),* 2007. doi:10.1109/IPDPS.2007.370480

[40] M. J. Rashti and A. Afsahi. "Assessing the ability of communication/computation overlap and communication progress in modern interconnects", in *Proceedings of the 15th IEEE Symposium on High-Performance Interconnects (HotI07),* 2007, pp. 117-124. doi:10.1109/ HOTI.2007.12

[41] M. J. Rashti and A. Afsahi, "A Speculative and adaptive MPI Rendezvous protocol over RDMA-enabled interconnects", *International Journal of Parallel Programming,* vol. 37, No. 2, April 2009, pp 223-246. doi:10.1007/s10766-009-0094-9

[42] RDMA Consortium. iWARP protocol specification. Available: http://www.rdmaconsortium.org

[43] C. B. Reardon, A. D. George and C. T. Cole, "Comparative performance analysis of RDMA-enhanced Ethernet", in *Workshop on High-Performance Interconnects for Distributed Computing*, 2005.

[44] H. Shah, J. Pinkerton, R. Recio and P. Culley. Direct data placement over reliable transports (version 1.0). RDMA Consortium, Available: http://www.rdmaconsortium.org/home/draft-shah-iwarp-ddp-v1.0.pdf

[45] J. C. Sancho, K. J. Barker, D. J. Kerbyson and K. Davis, "Quantifying the potential benefit of overlapping communication and computation in large-scale scientific applications," in *Proceedings of the 2006 ACM/IEEE Conference on Supercomputing (SC '06)*, 2006, pp. 125. doi:10.1145/1188455.1188585

[46] Sandia National Laboratories, "LAMMPS Molecular Dynamics Simulator," Available: http://lammps.sandia.gov/

[47] G. Shet, P. Sadayappan, D. E. Bernholdt, J. Nieplocha and V. Tipparaju, "A performance instrumentation framework to characterize computation-communication overlap in message-passing systems," in *Proceedings of the IEEE International Conference on Cluster Computing (IEEE Cluster),* 2006, pp. 1-12. doi:10.1109/CLUSTR.2006.311887

[48] S. Sur, H. Jin, L. Chai and D. K. Panda, "RDMA read based rendezvous protocol for MPI over InfiniBand: Design alternatives and benefits," in *Proceedings of the Eleventh ACM SIGPLAN Symposium on Principles and Practice of Parallel Programming (PPoPP '06),* 2006, pp. 32-39. doi:10.1145/1122971.1122978

[49] K. D. Underwood and R. Brightwell, "The impact of MPI queue usage on message latency," in *Proceedings of the International Conference on Parallel Processing (ICPP'04),* 2004, pp. 152-160. doi:10.1109/ICPP.2004.1327915

[50] Virtual Protocol Interconnect (VPI), Whitepaper, Mellanox Technologies, Inc., Available: http://www.mellanox.com/pdf/prod_architecture/Virtual_Protocol_Interconnect_VPI.pdf

[51] R. Zamani, Y. Qian and A. Afsahi, "An evaluation of the Myrinet/GM2 two-port networks," in *Proceedings of the 29th Annual IEEE International Conference on Local Computer Networks (LCN'04),* 2004, pp. 734-742. doi:10.1109/LCN.2004.20

In: Cluster Computing and Multi-Hop Network Research
Editors: C. Jimenez and M. Ortego, pp. 35-54
ISBN: 978-1-60876-186-9

Chapter 2

REALISTIC SIMULATION OF VEHICULAR AD-HOC NETWORKS

Mari Carmen Domingo*
Barcelona Tech University (UPC), Barcelona, Spain

Abstract

VANETS (Vehicular Ad-hoc Networks) are a form of MANETs (Mobile Ad-hoc Networks) that provide communication between vehicles and a fixed equipment or among nearby vehicles. VANET communication is a big challenge due to mobility issues (mobility depends on drivers' plans but at the same time drivers react to changing road conditions). Specific protocols designed to support comfort and traffic safety applications should be developed; they must be able to adapt to mobile scenarios. Since it is very difficult to validate the implementation of these protocols in a real world environment, a common way of evaluation performance is through simulations.

In the present paper we focus on the simulation of VANETs in the most possible realistic environments. It is necessary to integrate traffic and network simulation models because simulations based on simple mobility models do not reflect the reality of VANET scenarios. For this purpose, the most important current tools for the traffic simulation in a transportation environment are introduced. Afterwards, the open source micro-traffic simulator SUMO (Simulation of Urban Mobility), the popular tool for the simulation and evaluation of network protocols NS-2 (Network Simulator) and an efficient tool to rapidly generate realistic mobility models named MOVE (Mobility Model Generator for Vehicular Networks), have been selected. Realistic simulations in two typical VANET scenarios have been conducted: a highway and the streets of a city (one characteristic neighborhood of Barcelona (Eixample) in Spain). DYMO has been selected as a simple routing protocol that succeeds AODV and is very appropriate for mobile environments. Realistic simulations have been carried out with different number of vehicles, number of vehicles acting as sources, a variety of vehicles' speeds, several TCP and UDP flows, etc. The goodput and the impact of traffic collision in the proposed scenarios have also been analyzed.

Index Terms: VANET, traffic simulation, network simulation.

* E-mail address: cdomingo@ entel.upc.edu. Telematics Eng. Dep., Barcelona Tech University (UPC), Esteve Terradas, 7. 08860 Castelldefels (Barcelona), Spain.

Introduction

VANETS (Vehicular Ad-hoc Networks) are a form of MANETs (Mobile Ad-hoc Networks) that provide communication among nearby vehicles or between vehicles and a fixed equipment [1]. Vehicle-to-vehicle communication refers to the exchange of some valuable traffic information between vehicles, such as roadway conditions, location of traffic accidents, weather information, etc. Vehicle-to-roadside communication [2] refers to the message transmission from fixed roadside nodes to vehicular nodes (i.e. transmission of the up-to-date status of traffic signals and entertainment services) or to the message transmission from mobile vehicular nodes to fixed roadside nodes (i.e. emergency vehicles such as ambulances or fire engines can transmit emergency signals to change the status of the traffic lights in order to arrive quicker and safer to the destination).

VANET communication is a big challenge due to mobility issues (mobility depends on drivers' plans but at the same time drivers react to changing road conditions). Specific protocols designed to support comfort and traffic safety applications in these networks should be developed; they must be able to adapt to mobile scenarios. Since it is very difficult to validate the implementation of these protocols in a real world environment, a common way of evaluation performance is through simulations.

In the present paper we focus on the simulation of VANETs in the most possible realistic environments. It is necessary to integrate traffic and network simulation models because simulations based on simple mobility models do not reflect the reality of VANET scenarios. For this purpose, the most important current tools for the traffic simulation in a transportation environment are introduced. Afterwards, the open source micro-traffic simulator SUMO (Simulation of Urban Mobility), the popular tool for the simulation and evaluation of network protocols NS-2 (Network Simulator) and an efficient tool to rapidly generate realistic mobility models named MOVE (Mobility Model Generator for Vehicular Networks), have been selected. Realistic simulations in two typical VANET scenarios have been conducted: a highway and the streets of a city (one characteristic neighborhood of Barcelona (Eixample) in Spain). DYMO has been selected as a simple routing protocol that succeeds AODV and is very appropriate for mobile environments. Realistic simulations have been carried out with different number of vehicles, number of vehicles acting as sources, a variety of vehicles' speeds, several TCP and UDP flows, etc. The goodput and the impact of traffic collision in the proposed scenarios have also been analyzed.

The paper is structured as follows: Next, we introduce the main components for mobility model generation in VANETs. Subsequently, related work about traffic simulation is discussed. Later, the network simulati and interface tool are described. Afterwards, we give details about our medium access scheme and routing protocol. Later on, our simulation scenarios, parameters and simulation results are shown. Finally, the paper is concluded.

Mobility Concept

One of the greatest challenges for the development of VANETs is related to the development of realistic simulation models. It is extremely important to choose mobility models that reflect the real behavior of vehicles. Many surveys related to node mobility in wireless networks have been introduced in literature [3][4][5][6] and [7]. However, it is really difficult to apply them to VANETs because they are based on random mobility and in real life non-random patterns can be experienced like, for example, preferences for certain destinations. Therefore, many components have to be taken into account, related not only to the space domain of vehicular mobility (motion constraints due to the presence of obstacles, attraction points, etc) but also to the time constraints (different patterns can be observed at different times of the day). Figure 1 shows the main components for mobility model generation [8].

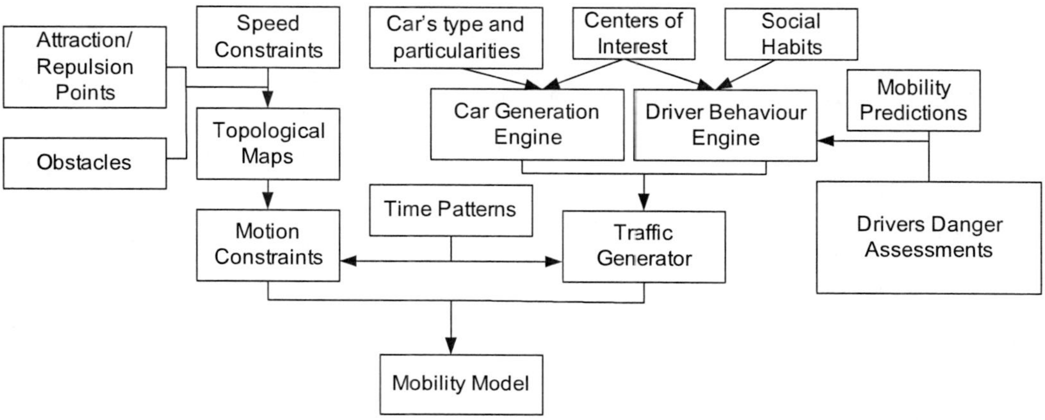

Figure 1. Map of mobility model generation for inter-vehicle communications.

Vehicular mobility models can be macroscopic or microscopic. In macroscopic models we focus on traffic flows and motion constraints such as roads, streets, crossroads and traffic lights. In microscopic approaches the movement of every single vehicle of the street and its behavior with respect to other vehicles are analyzed; for this reason, microscopic approaches are more suitable for VANET applications.

For the development of a simulator for VANET applications four concepts are essential: vehicle movement, wireless communication network, application behavior and the driver reactions according to information given by the application. The relationships between these concepts are illustrated in Figure 2.

Basically, two components are necessary to define the vehicle movement and network characteristics: a traffic generator and a network simulator. The vehicles' movement is reproduced with a traffic simulator whereas the network simulator imitates the behavior of the wireless communication network. The traffic simulator defines different types of cars and models their interactions with the environment such as overtaking, traffic jam, preferred paths, etc. It also defines the start and destination point for the vehicles, speed, movement, etc. The network simulator evaluates the nodes' position and connectivity patterns sent by the traffic generator and simulates the wireless communication defining the medium access, channel errors, routing protocol, security issues, etc. The application can be implemented as an additional module or as part of the network simulator. The driver behavior can be integrated into this application implementation, simplifying our simulation environment. The traffic generator and network simulator have to run simultaneously and exchange synchronization data continuously, as it is shown in Figure 3.

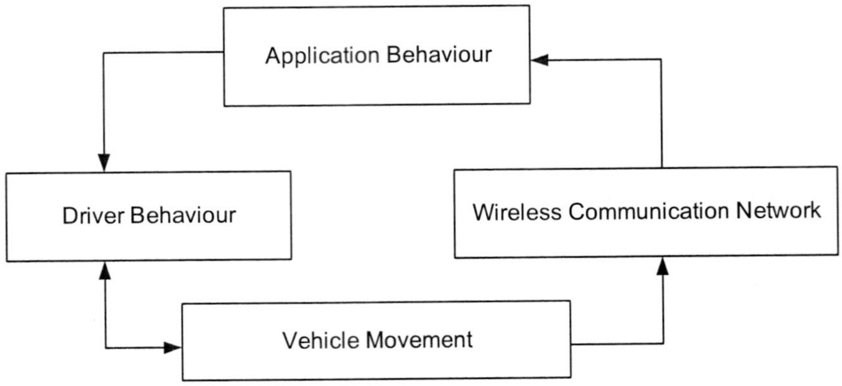

Figure 2. Most significant parameters for a traffic generator.

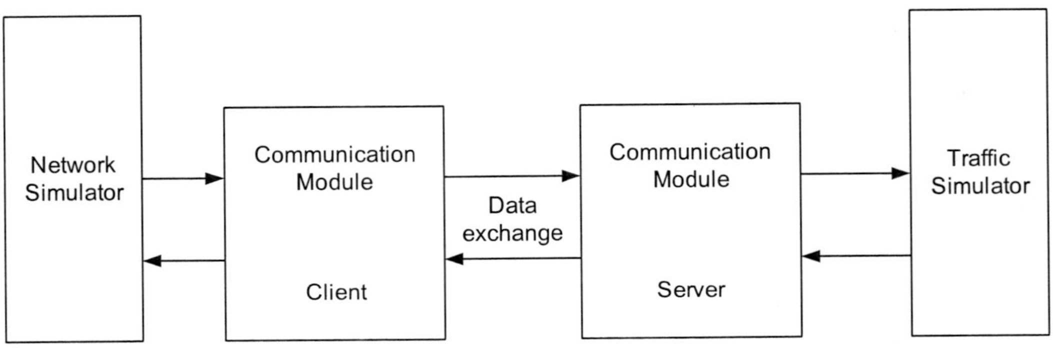

Figure 3. Coupling of network and traffic simulator.

Traffic Simulation

Many VANET-specific simulation tools have been developed by the research community to improve the understanding and analysis of VANETs [9][10] [11][12][13][14][15]. We introduce three popular simulation tools commonly used by the researchers: GrooveSim [14], CARISMA [15] and SUMO [16]. The first one tries to integrate traffic and network simulators, whereas the second and third ones are traffic simulators that generate mobility traces for NS-2 [17].

GrooveSim [14] is a vehicular network simulator that models inter-vehicular communication within a real street map-based topography. This simulator operates with traffic networks with thousands of vehicles and makes use of visual playback of driving logs (record of data related to the simulation). In the hybrid simulation mode real and simulated vehicles interact together. It uses GrooveNet, a diffusion-based geographic broadcast routing protocol.

CARISMA is also a traffic simulator that works with real topographical environments imported form digital map sources [15]. The main characteristics of CARISMA are the following: It can simulate mainly inner-city scenarios with hundreds to several thousand vehicles, it only supports two lanes per road, all streets have the same traffic capacity and priority, it assumes that buildings are located along all streets, it updates the vehicles positions and directions every second, it implements and evaluates different route and destination selection mechanisms and it computes connectivity information periodically. Besides, it is able to compute vehicle routes during simulation time. Vehicles move according to the Stefan Krauβ model [18].

Application	Application Name (Windows)	Application Name (Linux/UNIX)	Description
NETCONVERT	netconvert.exe	sumo-netconvert	A network converter/importer
NETGEN	netgen.exe	sumo-netgen	A generator of abstract networks
DFROUTER	dfrouter.exe	sumo-dfrouter	A router using detector flows
DUAROUTER	duarouter.exe	sumo-durarouter	A router for dynamic user assignment
JTRROUTER	jtrrouter.exe	sumo-jtrrouter	A router using junction turning ratios
SUMO	sumo.exe	sumo	The microscopic simulation
GUISIM	guisim.exe	sumo-guisim	The gui-version of the microscopic simulation
POLYCONVERT	polyconvert.exe	sumo-polyconvert	A tool for importing polygons from other formats
other	---	---	---

Figure 4. Most important applications of SUMO.

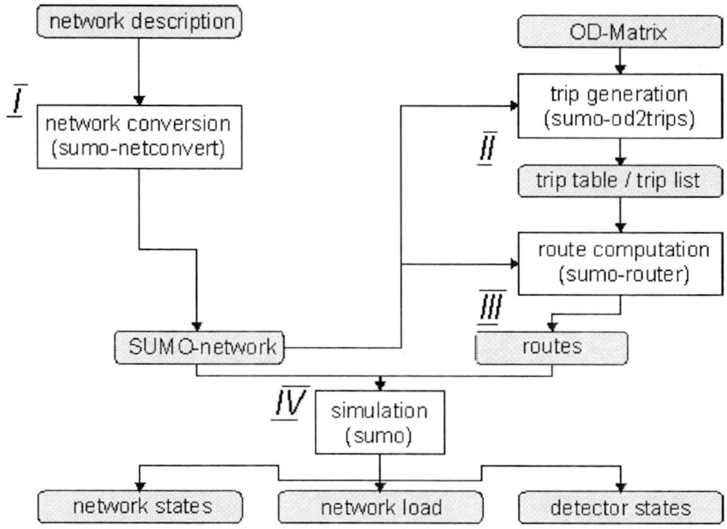

Figure 5. SUMO simulation (in blue data used, in white applications in use).

SUMO (Simulation of Urban Mobility) [16] is an open source, highly portable, microscopic road traffic simulation package designed to handle large road networks. It is developed by employees at the Deutsche Gesellschaft für Luft- und Raumfahrt (DLR). It is a space continuous and time discrete microscopic road traffic simulator. Space-continuous means that each vehicle has a certain position described by a floating-point number and the location of each vehicle can be determined at any time of the simulation. In SUMO the movement of every single vehicle is determined according to the Stefan Krauβ model [18] and it depends on the vehicle's physical abilities and the driver's controlling behavior. With SUMO, the routes for the vehicles are computed before the simulation starts; during the simulation, the vehicles arrive to their destinations using the shortest routes; as a result, the arterial roads can be congested. SUMO solves the problem using the dynamic user assignment approach developed by Christian Gawron. Since every vehicle knows its route, a large amount of data is necessary to describe the nodes' movements and the required memory to store the routes without compression can grow up to several MBs. This aspect should be improved in the future. The tasks of SUMO can be divided is small programs that interact together (see Figure 4 and Figure 5). More information related to SUMO can be found in the website [19].

Among the three presented traffic simulators, SUMO has been chosen due to its open architecture, the compatibility with existing map formats, the public software license that enables everybody to contribute to its improvement and the use of dynamic routing.

Network Simulation and Interface Tool

Network simulation is a technique where a program models the behavior of the network using mathematical formulas. NS-2 (Network Simulator) [17] has been selected as a widely accepted network simulator tool in the research community. It is open source and can work together with other traffic simulators such as SUMO. NS-2 can simulate different wireless scenarios, propagation models, medium access protocols, routing schemes, etc.

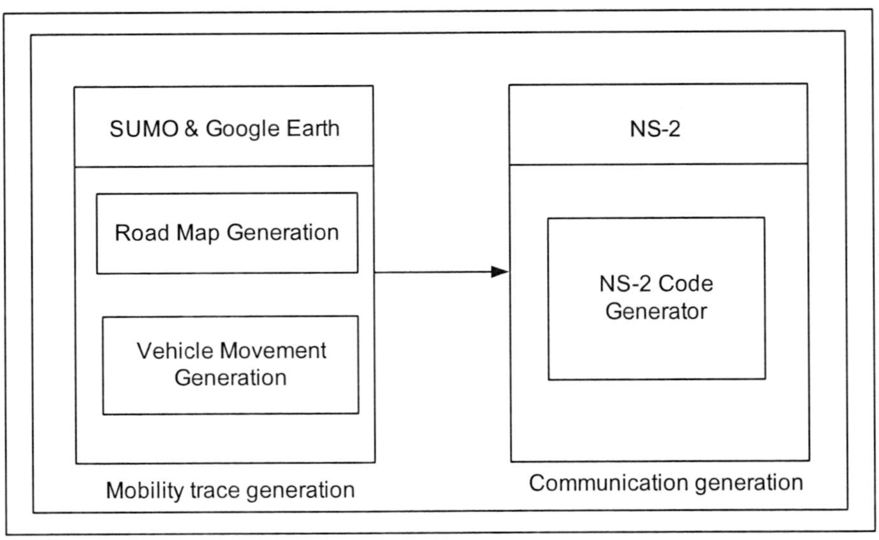

Figure 6. MOVE architecture.

MOVE (Mobility Model Generator for Vehicular Networks) [16] is the software implemented in Java that enables to communicate NS-2 and SUMO, making possible the simulation of VANETS. It allows the user to quickly generate realistic simulation scenarios without writing simulation scripts and without the need to learn about the internal details of the simulator. MOVE is currently implemented in Java and developed with the idea of enabling the rapid generation of realistic mobility models for VANETs. MOVE can incorporate realistic road maps into the simulation or interface with real world map databases such as TIGER (Topologically Integrated GEographic Encoding and Referencing) [20] and

Google Earth. It consists of a Map Editor to create the road topology and a Vehicle Movement Editor to specify the trips of the vehicles and the route that each vehicle will take for one particular trip (see Figure 6).

Medium Access Control and Routing Protocol

Since our objective was to simulate VANETs in a realistic way, the ideal case would be to use protocols specifically designed for this type of networks such as IEEE 802.11p as medium access scheme. IEEE 802.11p is a draft amendment to the IEEE 802.11 standard to add wireless access in the vehicular environment. The 802.11p Task Group is still active and the work plan predictions are to publish the IEEE 802.11p amendment in November 2010; thus, it is not already available. On the other hand, many researchers have successfully used the IEEE 802.11b standard operating in DCF (Distributed Coordination Function) mode to simulate VANETs and this will be the standard used for our simulations.

As routing protocol we have selected DYMO (Dynamic On-demand MANET routing protocol) [21], a protocol with similar traits compared to AODV (Ad Hoc On Demand Distance Vector) but modified to work properly in mobile environments as VANET scenarios.

Now we summarize the main characteristics related to our medium access and routing protocol.

IEEE 802.11

The basic scheme for DCF (Distributed Coordination Function) is the Carrier Sense Multiple Access with Collision Avoidance (CSMA/CA) [22]. When a station wants to send a packet, it must first sense the medium. If the channel is sensed idle for a time interval equal or greater than a Distributed Inter-Frame Space (DIFS), then the station can send the packet; otherwise the station waits a random time after the channel being idle for DIFS, named the backoff time, and then it sends the packet. The backoff time interval can be computed as follows:

$$T_{backoff} = Rand(0, CW) \cdot T_{slot} \tag{1}$$

where T_{slot} represents the time slot selected by the physical layer. This formula means that the node chooses a backoff time equal to a random number of time slots between 0 and a contention window CW. The backoff interval is used to initialize a backoff timer; which will be decreased for each station while the medium is idle and will be frozen when the transmission from another station is detected. If the medium remains idle for a period greater than DIFS, the backoff timer is periodically decremented by one for every time slot. When the backoff timer expires the station accesses immediately the medium. If two or more stations start transmission at the same time slot, a collision occurs.

To reduce the probability of consecutive collisions, the CW is doubled after each unsuccessful transmission attempt until a maximum value (CW_{max}) is reached. After a successful transmission, the CW is reset to CW_{min}.

VANETs are distributed ad hoc networks and the PCF (Point Coordination Function) mode of IEEE 802.11 is centralized; for this reason the distributed DCF mode is more appropriate.

DYMO

Dynamic MANET On-demand (DYMO) routing protocol is reactive, which means that routes are set up only when needed (when two or more nodes wish to communicate). The two basic operations of DYMO are route discovery and route maintenance.

When a node needs to send a packet to a destination and the route is available in its routing table, it uses the stored route. Otherwise, a Route Request message is flooded in the network (RREQ) using broadcast in the route discovery process. During this hop-by-hop dissemination process, each intermediate node records a route to the source node. When the packet reaches its destination, a Route Reply message is sent back containing the discovered accumulated path. Each intermediate node that receives the RREP creates a route to the target, and then the RREP is unicasted hop-by-hop toward the source node. When the source node receives the RREP, routes have been established between the source node and the destination node in both directions.

Route maintenance is a process designed to be aware of topology changes and update the routing information after route establishment. It consists of two operations. Nodes continuously monitor de active links and extend route lifetimes when forwarding data packets. The source node receives a notification when a data packet is received for forwarding and a route for the destination is not known or is unavailable (it does not work). A Route Error (RERR) is sent toward the packet source to indicate the current route to a particular destination is invalid or missing. When the source node receives the RERR, it deletes the route. If the source node later receives a packet to be forwarded to the same destination, it will need to perform route discovery again for that destination.

DYMO uses sequence numbers to achieve loop freedom.

Simulation Scenarios

In the next section we introduce the different scenarios that consist of a highway and a typical urban area. Different cases have been analyzed for each scenario using SUMO as traffic simulator, NS-2 as network simulator and MOVE as interface tool. The simulation parameters that characterize our simulations are described as well.

Simulation Scenarios

We have selected two typical VANET scenarios: a highway and the streets of a city (we have modeled one characteristic neighborhood of Barcelona (Eixample) in Spain). Now we will describe them.

Highway

As first scenario we consider a highway where vehicles move in the same direction using four lanes. The average width of each lane is 2.7 m (we consider that a vehicle fits in one lane). The vehicles are distributed among the four existing lanes, with exception of the simulation case with 20 vehicles, where these vehicles only drive in the two central lanes (10 vehicles in each lane). We assume that the vehicles travel grouped and separated by a minimum safety distance between each other. Besides, they have a constant speed, since traffic in a highway for large and clear paths tends to have these characteristics. Our objective was to evaluate the goodput per source and the impact of speed on it (among other parameters).

Urban Area: Eixample of Barcelona

As second scenario we have recreated 49 blocks of Eixample, a famous neighborhood in Barcelona (Spain) with two-way streets (two lanes). The total reproduced extension was 6,531 m^2 =133.3 m^2 (block extension)*49 (blocks), an extension equivalent to 6 and a half football fields.

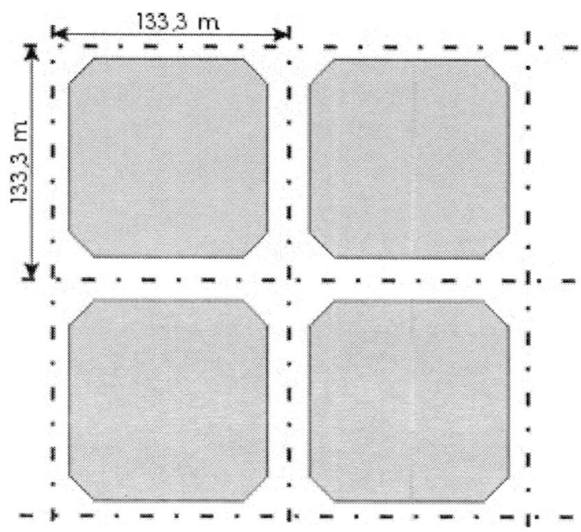

Figure 7. Basic urban sector of Eixample in Barcelona.

The selected scenario has been reproduced adapting the measurements of the Eixample blocks in Barcelona into a map with the tools offered by SUMO. The typical traffic conditions of urban areas have been incorporated to make more realistic simulations: Vehicles drive around the blocks dynamically, each vehicle moves at a different speed in a different way, vehicles suffer different accelerations and decelerations, etc.

Simulation Cases

In the simulations, we evaluate the behavior of TCP and UDP traffic and different applications:

TCP traffic:

Vehicle-to-vehicle communications to interchange information about the road state, weather, etc. with the aim of increasing comfort during the trip as well as avoiding congestion. This information will be displayed to the passengers of the vehicle on a screen.

UDP traffic:

Vehicle-to-vehicle communications where multimedia traffic is exchanged, e.g. videos clips related to a stretch of the road in bad conditions, etc. to complement the previously given information.

The selected evaluation cases due to its importance and interest will be:

Case 1: TCP/UDP traffic as a function of speed, number of vehicles and percentage of sources in a highway.

- Case 1.1 TCP traffic with 20% sources
- Case 1.2 TCP traffic with 80% sources
- Case 1.3 UDP traffic with 20% sources
- Case 1.4 UDP traffic with 80% sources

Case 2: TCP/UDP traffic and impact of a traffic accident in the communication on a highway as a function of the percentage of sources.

- Case 2.1 TCP traffic
- Case 2.2 UDP traffic

Case 3: TCP/UDP traffic and the impact of a traffic accident on an urban area as a function of the percentage of sources.

- Case 3.1 TCP traffic
- Case 3.2 UDP traffic

Case 4: TCP/UDP traffic and impact on an urban area (Eixample of Barcelona) compared to the impact on a highway as a function of the percentage of sources.

- Case 4.1 TCP traffic
- Case 4.2 UDP traffic

Simulation Parameters

We will evaluate the impact of mobility models on the goodput using DYMO as routing protocol. We define the goodput per source, that is, the traffic that arrives successfully to the receivers as:

$$Goodput = \frac{Maximum\ Number\ of\ received\ Packets \times Packet\ Size}{Measurement\ \ Interval}$$

We have computed the goodput per source as follows:

$$\begin{matrix} Goodput \\ per\ Source \end{matrix} = \frac{Maximum\ Number\ of\ received\ Packets \times Packet\ Size}{Number\ of\ Sources \times Measurement\ Interval}$$

Each simulation lasts for 1050 seconds. We generate scenarios from 20 up to 200 vehicles (increases of 20 nodes) for Case 1 and for 20 vehicles for Cases 2, 3 and 4. The number of source nodes varies from 20% to 80% (increases of 10) for Cases 2, 3 and 4; for Case 1 the number of source nodes is 20% or 80%. Each source node is a CBR traffic source transmitting UDP packets of 512 bytes at a rate of 32 Kbps or a TCP Reno source transmitting TCP packets of 512 bytes with a window size of 32. The vehicles move at a speed from 60 to 140 Km/h (increases of 20) for Case 1 and 100 Km/h for Case 2. Vehicles for Case 3 move at speeds between 10 and 50 Km/h and they suffer accelerations and decelerations. Vehicles for Case 4 move in the same way as in Case 3 if they drive in an urban area and at 100 Km/h if they drive in a highway. The computation of the safety distance between vehicles will depend on the speed. SUMO computes the safety distance between vehicles using the Stefan Krauβ algorithm [18]. The following formula is an approximation to compute the safety distance (in meters):

$$Minimum\ safety\ distance = \frac{V^2}{C} \tag{2}$$

where V is the speed of the vehicle in Km/h and C is a constant equal to 170.

All nodes use 802.11b MAC operating at 2 Mbps. The transmission range is 250 m. The propagation model employed in the simulation is the log normal shadowing model. We use a path loss exponent 2.56 with standard deviation 4.0 based on real-world measurement data from inter-vehicle experiments carried in Sydney suburban area [23].

Simulation Results

In this section, we present our simulation results. We distinguish between two basic scenarios: highway and urban area.

Highway Scenario

First we have run simulations for traffic in a highway (Case 1). Figure 8 shows the goodput per source as a function of speed for TCP traffic with 20% sources (Case 1.1).

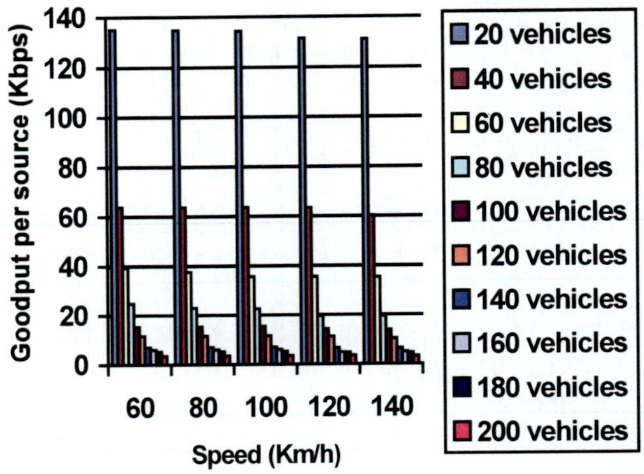

Figure 8. TCP traffic with 20% sources.

We observe that the goodput is decreased exponentially with an increasing number of vehicles. The goodput values are around 135 Kbps with 20 vehicles (best goodput) and around 3 Kbps with 200 vehicles (worst goodput). We also appreciate that the variations in speed do not really affect goodput, since the graphs have very similar values at different speeds. The reason is that in the proposed scenario all vehicles run with the same constant speed; therefore, with SUMO all vehicles in the scenario are distributed keeping the same distance between them, the minimum safety distance (with a maximum speed of 140 Km/h the safety distance will be 115.3 meters). This means that in all cases there is always communication between vehicles because 115.3 meters is lower than the transmission range (250 meters). The increase in the number of vehicles in the scenario does negatively affect goodput per source because a higher number of nodes (the number of intermediate nodes is increased) try to send packets, causing more congestion, collisions and retries to access the medium. Consequently, fewer packets arrive to their destinations successfully.

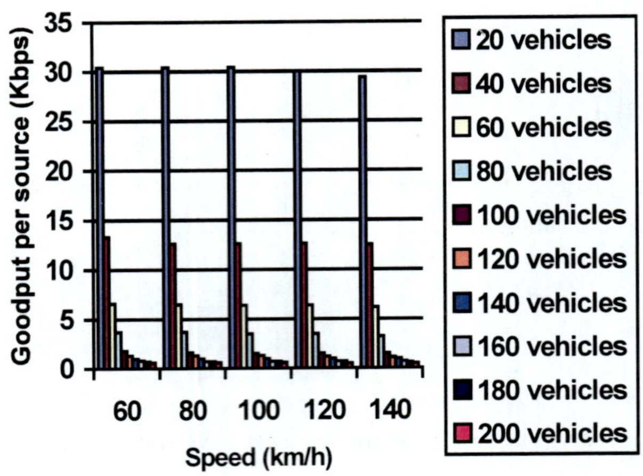

Figure 9. TCP traffic with 80% sources.

Figure 9 shows the goodput per source as a function of speed for TCP traffic with 80% sources (Case 1.2). We observe the same effects as in Figure 8. If the percentage of traffic sources is increased from 20% to 80%, the resulting goodput per source values are much lower because more nodes transmit packets and, as a result, the number of collisions is increased and the goodput is decreased. In Figure 8 goodput values with 20% sources between 135 and 3 Kbps are obtained, whereas with 80% sources the values vary between 30 and 0.5 Kbps.

Figure 10 shows the goodput per source as a function of speed for UDP traffic with 20% sources (Case 1.3). We observe the same effects as in Figure 8. Comparing TCP with UDP we notice that the goodput values per source are very similar in both scenarios; the goodput values for UDP are slightly smaller (between 131 Kbps and 3 Kbps instead of between 135 Kbps and 3 Kbps as in TCP), because UDP is connectionless (it sends data without ever establishing a connection) and it sends traffic at a constant rate without receiving any feedback about the packet reception. Consequently, a high number of packets are lost due to buffer overflow (this phenomenon has been observed in the simulations). On the contrary, TCP is connection-oriented (before one application process can begin to send data to another, the two processes must first "handshake" with each other, that is, they must send some preliminary packets to each other to establish the parameters of the ensuing data transfer) and offers reliable data transfer with retransmission of lost and corrupted data. TCP suffers from buffer overflow as well (in our simulations the buffer size is 50 packets) but, in this case, when packets are lost, they are later retransmitted and therefore, goodput values are increased in comparison with UDP. We conclude that UDP sends more packets but congests more the medium and, as a result, fewer packets arrive to their destinations compared to TCP, although initially a higher number of packets had been sent.

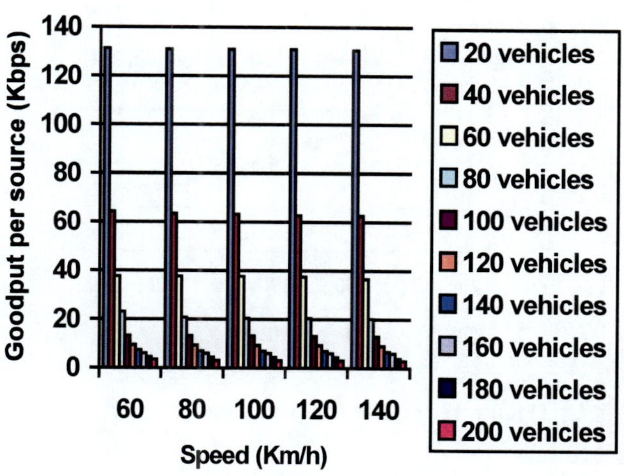

Figure 10. UDP traffic with 20% sources.

Figure 11 shows the goodput per source as a function of speed for UDP traffic with 80% traffic sources (Case 1.4). We can again observe the decrease of goodput due to an increase in the percentage of sources (between 133 Kbps and 3 Kbps with 20% sources and between 27 Kbps and 0.5 Kbps for 80% sources). We conclude that traffic (UDP or TCP) is not affected by the different speed of vehicles but diminishes when the percentage of sources is increased.

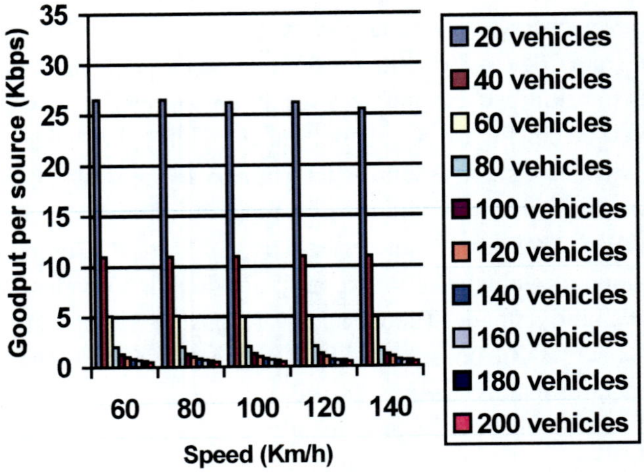

Figure 11. UDP traffic with 80% sources.

Now we have analyzed the impact of a traffic accident on a highway communication as a function of the percentage of sources. We assume that in the middle of the simulation two cars suffer a traffic accident and do not continue driving as the rest of the vehicles. Consequently, links fail and new routes are established for communication between nodes in this scenario.

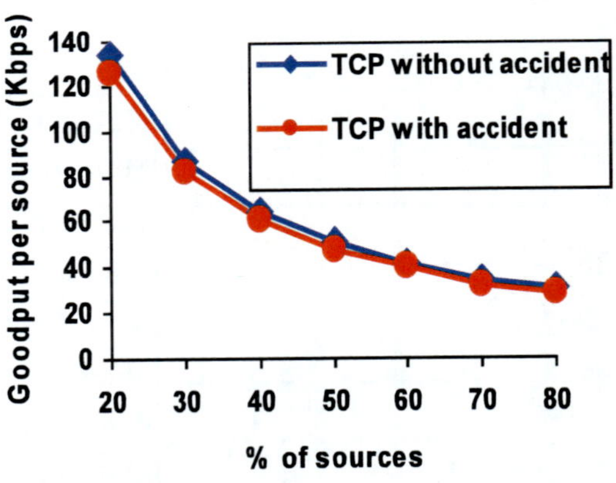

Figure 12. TCP traffic with/without accident.

We consider a scenario where 20 vehicles drive at a speed of 100 Km/h. Figure 12 shows the goodput per source for TCP traffic with and without a traffic accident as a function of the percentage of sources (Case 2.1). We observe that the goodput per source decreases exponentially when the percentage of traffic sources is increased. The impact of an accident on the communication between nodes is noticed because goodput is diminished from values between 134 Kbps and 30 Kbps for the case without accident to values between 126 Kbps and 28 Kbps for the case with accident. Accidents affect goodput because certain routes are broken down, packets are lost and reestablishment of new routes to send information is necessary.

Figure 13 shows the goodput per source for UDP traffic with and without a traffic accident as a function of the percentage of sources (Case 2.2). We can observe the same effects as previously described in Figure 12. We obtain values for UDP traffic between 131 Kbps and 26 Kbps without accident and between 119 Kbps and 22 Kbps with accident. The comparison between TCP and UDP values shows that they are similar (UDP values are in fact slightly lower due to the same reasons specified in the explanations of Figure 10).

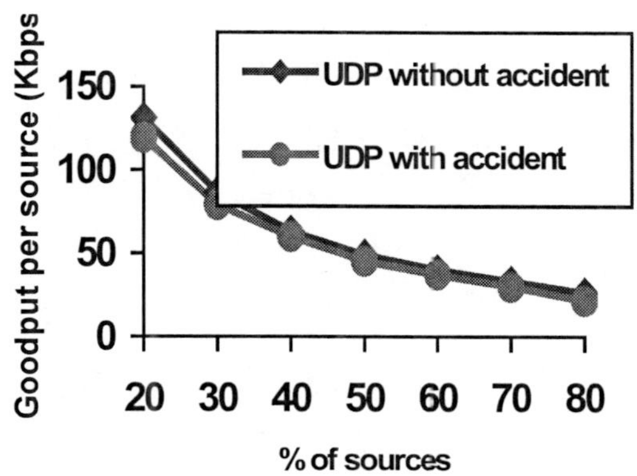

Figure 13. UDP traffic with/without accident.

Urban Area: Eixample of Barcelona

Now we will simulate the traffic of an urban area. Figure 14 shows the goodput per source for TCP traffic with and without a traffic accident as a function of the percentage of sources (Case 3.1). We observe that after the accident the goodput is diminished because after routes are broken down, packets are lost and new routes have to be reestablished to send information. Goodput values for TCP traffic vary between 130 and 29 Kbps without accident and between 124 and 27 Kbps with accident.

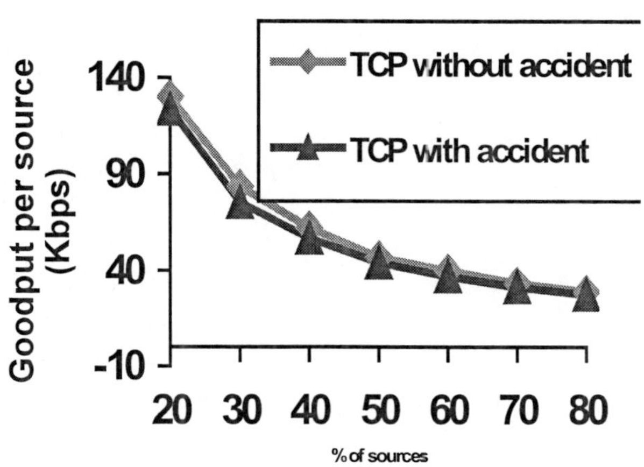

Figure 14. TCP traffic with/without accident.

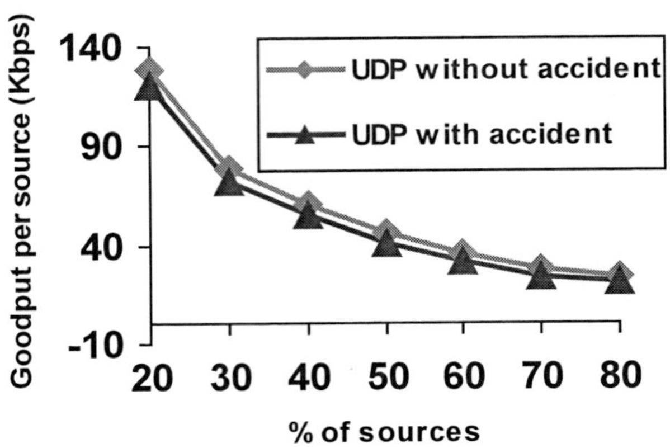

Figure 15. UDP traffic with/without accident.

Figure 15 shows the goodput per source for UDP traffic with and without a traffic accident as a function of the percentage of sources (Case 3.2). We observe the same effects as in Figure 14. Goodput values per source for UDP traffic between 129 and 24 Kbps without accident and between 121 and 22 Kbps with accident have been measured. TCP and UDP values are again very similar, although UDP values are slightly lower for already explained reasons (for TCP traffic between 130 and 29 Kbps without accident and between 124 and 27 Kbps with accident).

Comparison between Highway and Urban Area

Now we will compare the traffic of a highway where vehicles drive at a speed of 100 Km/h with the traffic of an urban area where vehicles drive at speeds between 10 and 50 Km/h (different vehicles drive at different speeds, they suffer accelerations, decelerations, etc.). We can compare both traffic types although the speeds of the vehicles are different because speed does not affect goodput, as we have previously seen in our simulation results. Figure 16 shows the goodput per source for TCP traffic as a function of the percentage of sources for the highway and urban area scenarios (Case 4.1). We can observe that in the urban area the goodput values are slightly lower than those in the highway (although all values are very close, between 134 Kbps and 30 Kbps in the highway and between 130 Kbps and 29 Kbps in Barcelona Eixample). The reason is that in the urban area each vehicle drives at a different speed (speeds vary between 10 and 50 Km/h) and suffers different accelerations and decelerations, whereas in the highway all vehicles run at a constant speed (100 Km/h); therefore, in the urban area distances are lower (in the highway the distance between vehicles has been computed as 58.8 meters) and vehicles can sometimes drive very close together as they run in Eixample and, if they want to communicate with each other, the probability of collision between packets and retrials is increased because many nodes will contend for medium access. Thus, goodput is diminished.

Figure 17 shows the goodput per source for UDP traffic as a function of the percentage of sources for the highway and urban area scenarios (Case 4.2). In this case we can notice the same effects as in Figure 16. The goodput values per source vary between 131 and 26 Kbps for highway and between 129 and 24 Kbps for Barcelona Eixample. Again, we obtain similar TCP and UDP values, whereas the UDP values are slightly smaller for already explained reasons (for TCP traffic they vary between 134 and 30 Kbps in the highway and between 130 and 29 Kbps in an urban area).

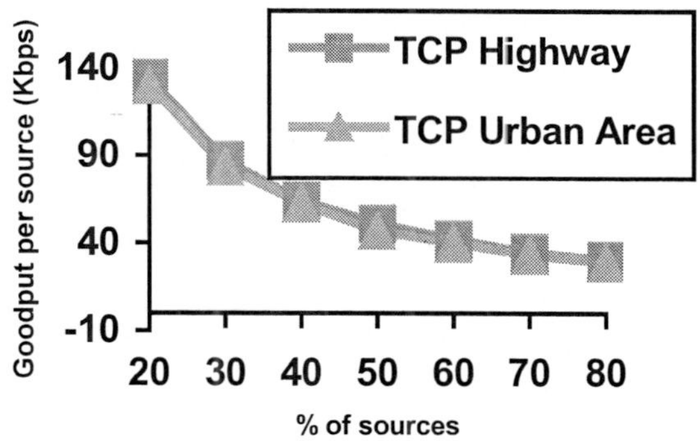

Figure 16. TCP traffic for highway and urban area.

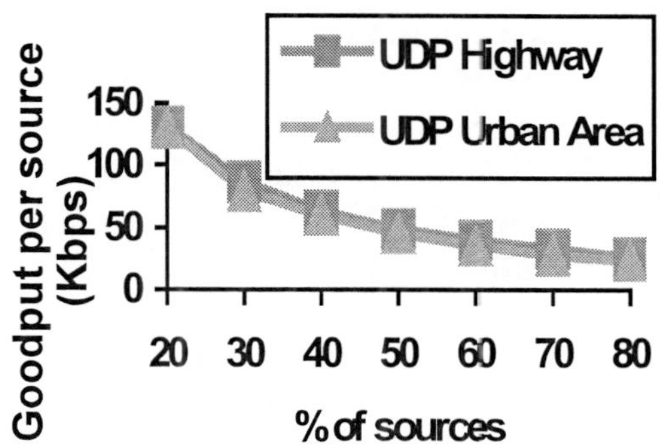

Figure 17. UDP traffic for highway and urban area.

Conclusion

In the present paper the most important current tools for the traffic simulation in a transportation environment have been introduced. The micro-traffic simulator SUMO, the network simulator NS-2 and the interface tool MOVE have been selected to generate realistic simulations of VANETs in two typical VANET scenarios: a highway and the streets of a city (one characteristic neighborhood of Barcelona (Eixample) in Spain). DYMO has been selected as a simple routing protocol. We have measured the goodput and the impact of a traffic accident in different scenarios varying the number of vehicles, number of sources, TCP or UDP flows, etc.

We conclude that vehicle-to-vehicle communication in VANETs works properly; we obtain goodput values per source for TCP traffic of 135 Kbps and for UDP traffic of 131 Kbps in the best cases.

Our simulation results show no impact of speed on goodput per source in a highway. The reason is that in this scenario all vehicles drive at the same constant speed and even at the highest speed (140 Km/h) it is possible to communicate vehicles because the minimum safety distance is 115.3 metes, a distance lower than the transmission range (250 meters). For higher speeds the minimum safety distance would be higher and for closer values to the transmission range some packets would be lost because they would not be able to reach the minimum power detection threshold.

The goodput per source diminishes exponentially in the highway as well as in urban areas due to:

- The increase in the number of sources in a scenario where the number of vehicles is kept constant.
- The increase in the number of vehicles in a scenario where the percentage of sources is kept constant.

In both cases goodput is diminished because more vehicles (source or intermediate nodes) will send packets. As a result, the number of collisions and retrials due to contention for medium access is increased and fewer packets reach their destinations than in the case where the number of sources or vehicles is lower.

We have also observed that goodput values for UDP traffic are slightly lower than those for TCP traffic. Throughput values are higher for UDP because this protocol is connectionless and sources send traffic without waiting for acknowledgements. But more packets are lost with UDP than with TCP, because TCP is connection-oriented and offers packet retransmission. TCP is more appropriate for applications that require reliability and it is shown in our simulations that the goodput per source is increased with this type of traffic. Although UDP sends more packets, the medium is more congested and, as a result, fewer packets arrive to their destinations.

Our simulation results show that with a traffic accident goodput is diminished because routes are broken down, packets are lost and route reestablishment is necessary to send again information.

Our simulation results also show that the goodput per source is lower in urban area scenarios than in the highway scenarios because in urban area scenarios vehicles drive at different speeds between 10 and 50 Km/h, suffer accelerations and decelerations and the minimum safety distance is lower than in the highway scenarios. In urban areas vehicles drive sometimes very close between each other and for this reason contention for medium access between nodes is increased, collisions and retrials are increased as well and goodput is diminished. Therefore, we conclude that node density will be a decisive parameter in the results, because an increase in the number of nodes causes congestion and affects goodput. There is no impact of speed on goodput per source in urban areas.

The simulations in urban areas could be made more realistic if not only the streets' topography but also the existence of buildings is described, because the presence of obstacles makes radio communication more difficult due to signal bouncing, fading, etc. As future work we propose to develop some software to incorporate these effects in our simulations.

Acknowledgment

This work was supported by the MEC (the Spanish Ministry of Education and Science) under the project TSI2007-66637-C02-01, which is partially funded by FEDER.

References

[1] Li, C.-T; Hwang, M.-S; Chu, Y.-P. A secure and efficient communication scheme with authenticated key establishment and privacy preserving for vehicular ad hoc networks. *Computer Communications*, vol. 31, issue 12, pp. 2803-2814, 2008.

[2] Dotzer, F.; Kohlmayer, F.; Kosch, T.; Strassberger, M. Secure communication for intersection assistance. In Proceedings of the *2nd International Workshop on Intelligent Transportation*, Hamburg, Germany, 2005.

[3] Camp, T.; Boleng, J.; Davies, V. A survey of mobility models for ad hoc network research. In *Wireless Comm. and Mobile Computing (WCMC)*, vol. 2, no. 5, pp. 483-502, 2002.

[4] Bettstetter, C. Smooth is better than sharp: A random mobility model for simulation of wireless networks. In Proc. of the *Fourth ACM International Workshop on Modeling, Analysis and Simulation of Wireless and Mobile Systems*, pp. 19-27, 2001.

[5] Bai, F.; Helmy, A. A Survey of Mobility Modeling and Analysis in Wireless Ad hoc Networks. *Book Chapter in the book on Wireless Ad Hoc and Sensor Networks*, Kluwer Academic Publishers, June 2004.

[6] Le Boudec, J.-Y.; Vojnovic, M., Perfect Simulation and Stationarity of a Class of Mobility Models. In *Proc. of the Infocom'05*, USA, 2005.

[7] Broch, J.; Maltz, D. A.; Johnson, D. B.; Hu, Y.-C.; Jetcheva, J. G. A Performance Comparison of Multi-Hop Wireless Ad Hoc Network Routing Protocols. In Proc. of the *Fourth Annual ACM/IEEE International Conference on Mobile Computing and Networking (MobiCom'98)*, pages 85-97. ACM/IEEE, October 1998.

[8] Härri, J; Filali, F.; Bonnet, C. A framework for Mobility Models Generation and its Application to Inter-Vehicular Networks. *Wireless Networks, Communications and Mobile Computing*, 2005 International Conference on, vol. 1, pp. 42- 47, June 2005.

[9] Choffnes, D.; Bustamante, F. An integrated mobility and traffic model for vehicular wireless networks. In *VANET '05: Proceedings of the 2nd ACM international workshop on Vehicular ad hoc networks,* New York, NY, USA: ACM Press, 2005, pp. 69–78.'05.

[10] Härri, J.; Fiore, M.; Fethi, F.; Bonnet, C. VanetMobiSim: generating realistic mobility patterns for VANETs. In Proc. of the *3rd ACM International Workshop on Vehicular Ad Hoc Networks (VANET'06),* September 29, 2005, Los Angeles, USA.

[11] Lochert, C.; Barthels, A.; Cervantes, A.; Mauve, M.; Caliskan, M. Multiple simulator interlinking environment for IVC. In 2nd *ACM International Workshop on Vehicular Ad hoc Networks (VANET 2005),* Poster Session. Cologne, Germany: ACM Press, September 2005, pp.87–88.

[12] Wang, S. Y.; Chou, C. L.; Chiu, Y. H.; Tseng, Y. S.; Hsu, M. S.; Cheng, Y. W.; Liu, W. L.; Ho, T. W. NCTUns 4.0: An Integrated Simulation Platform for Vehicular Traffic, Communication, and Network Researches. In 1st *IEEE International Symposium on*

Wireless Vehicular Communications (WiVec 2007), Baltimore, MD, USA, October 2007.

[13] Naumov, V.; Baumann, R.; Gross, T. An evaluation of inter-vehicle ad hoc networks based on realistic vehicular traces. In Proceedings of the 7th *ACM international symposium on Mobile ad hoc networking and computing (MobiHoc'06)*, Florence, Italy, May 2006, pp. 108–119.

[14] Mangharam, R.; Weller, D. S.; Stancil, D. D.; Rajkumar, R. GrooveSim: *A Topography-Accurate Simulator for Geographic Routing in Vehicular Networks*. ACM Mobicom/VANET, Sept. 2005.

[15] Eichler, S.; Ostermaier, B.; Schroth, C.; Kosch, T. *Simulation of Car-to-Car Messaging: Analyzing the Impact on Road Traffic. MASCOTS'05*, 507-510, 2005.

[16] Karnadi, F.; Mo, Z.; Lan, K.-C. *Rapid Generation of Realistic Mobility Models for VANET*. WCNC'07, March 2007.

[17] NS-2, *Network Simulator*: *http://www.isi.edu/nsnam/ns/*.

[18] Krauß, S. *Microscopic Modeling of Traffic Flow: Investigation of Collision Free Vehicle Dynamics*. PhD thesis, Universität zu Köln, 1998.

[19] SUMO, *http://sumo.sourceforge.net/*

[20] TIGER *(Topologically Integrated GEographic Encoding and Referencing)*, *http://www.census.gov/geo/www/tiger/*.

[21] Chakeres, I. D.; Perkins, C. E. Dynamic MANET On-demand Routing Protocol. *IETF Internet Draft*, draft-ietf-manet-dymo-12.txt, February 2008 (Work in Progress).

[22] Schiller, J. H. *Mobile Communications*. ed. Addison Wesley Professional, 2000.

[23] Karnadi, F.; Mo, Z.; Lan, K.-C. Rapid Generation of Realistic Mobility Models for VANET. 11th Annual *International Conference on Mobile Computing and Networking (MobiCom 2005)*, Poster Session, Cologne, Germany, August 2005.

In: Cluster Computing and Multi-Hop Network Research ISBN: 978-1-60876-186-9
Editors: C. Jimenez and M. Ortego, pp. 55-71

Chapter 3

COMPUTATIONAL METHODS FOR SPATIAL DISTRIBUTION MODELING OF ARTHROPODS

WenJun Zhang*
Research Institute of Entomology, School of Life Sciences, Sun Yat-sen University, Guangzhou 510275, China.

Abstract

Spatial distribution models, like Poisson, binomial, and negative binomial functions, etc., have no power to predict spatial distribution. This study aimed to develop some models, and use these models to describe or predict spatial distribution of arthropods. Arthropod investigation was conducted on the grassland. An artificial neural network and a partial differential equation were developed for modeling field distribution of arthropods. Several models were compared and evaluated for their effectiveness in modeling spatial distribution.

Most arthropods on the grassland were insects. The most abundant orders were found to be Homoptera, Orthoptera, and Hymenoptera. The results demonstrated that artificial neural network exhibited an excellent simulation performance. Simulated spatial distribution was perfectly in accordant with the observed one. Overall neural network performed better in the case of lower total abundance on the grassland. Response surface model may better fit the spatial distribution of arthropods but the simulation performance of this model was lower than neural network. Cross validation revealed that neural network yielded much better performance than response surface model and spline function in predicting spatial distribution of arthropods. Confidence interval of modeled abundance could be obtained using randomized submission of quadrate sequences in the neural network modeling. A method for obtaining parametrical values in the partial differential equation was presented. The partial differential equation developed in this study may be used to make extrapolation on spatial distribution of arthropods. It is concluded that artificial neural network is a powerful model to describe the spatial distribution of arthropods.

Key words: Artificial neural network; Partial differential equation; Response surface model; Spline function; Arthropods; Spatial distribution; Modeling

* E-mail address: zhwj@mail.sysu.edu.cn

1. Introduction

Arthropods account for 90% of global species. Biomass of arthropods reaches 1,000kg /ha in a temperate grassland, lower than plant (20,000kg/ha) and microorganisms (7,000kg /ha), but higher than mammals (1.2kg/ha), birds (0.3kg/ha), et al (Pimental et al., 1992). Arthropods control the structures and functions of ecosystems (Wilson, 1987). Moreover, arthropods have been considered as the sensitive indicators of environment health (Brown, 1991; Kremen et al., 1993).

Spatial distribution means the distribution of animal or plant individuals in a space, particularly on ground. A lot of probabilistic distribution functions and aggregation indices were developed or used to describe spatial distribution of individuals (Krebs, 1989). However, the spatial information is not available from those models. Due to the lack of theoretical background it is also hard to construct a mechanistic model that calculates individual distribution from spatial information. Questions on spatial distribution are therefore data-driven (Schultz and Wieland, 1997). The relationship between individual distribution and spatial information is usually a nonlinear relationship (Pastor-Barcenas et al., 2005).

Artificial neural networks are known to be flexible and adptable function approximators for nonlinear relationships (Cereghino et al., 2001; Marchant and Onyango, 2003; Acharya et al., 2006; Filippi and Jensen, 2006; Nour et al., 2006; Zhang, 2007a,b; Zhang and Barrion, 2006; Zhang et al., 2007; Zhang et al., 2008). They can offer the advantages of simplified and more automated model synthesis and analytical input-output models (Abdel-Aal, 2004; Tan et al., 2006). A large number of studies were reported concerning applications of neural networks. For examples, they are considered to be more effective in time series prediction than previous procedures based on dynamical system theory (Ballester et al., 2002). They are used in the forecast of short and middle long-term concentration levels (Viotti et al., 2002), subsurface modeling (Almasri and Kaluarachchi, 2005), modeling hourly temperature with the alternative abductive networks (Abdel-Aal, 2004), modeling sediment transfer (Abrahart and White, 2001), and subsurface drain outflow and nitrate-nitrogen concentration in tile effluent and surface ozone (Sharma et al., 2003; Pastor-Barcenas, et al., 2005), and estimation of endoparasitic load using morphological descriptors (Loot et al., 2002), the uses of BP to describe nitrogen dioxide dispersion (Nagendra and Khare, 2006), and for reservoir eutrophication prediction (Kuo et al., 2007). Empirical models regained popularity in recent years due to the complexity and nonlinearity of ecosystems (Tan et al., 2006). Various conventional models, including empirical models, were thus used to compare simulation performances between neural networks and these models. For instance, it demonstrated that neural network was superior to linear models, generalized additive models, and classification and regression trees (Moisen and Frescino, 2002). Neural network was proved to outperform other models like multiple regression, logistic regression, and multiple discriminant model in predicting the number of salmonids and community composition (McKenna, 2005; Olden et al., 2006). In the research areas of arthropods or related taxonomic groups, neural networks have been used to make simulation and prediction. A stream classification based on characteristic invertebrate species assemblages was also satisfactorily conducted using self-organizing map neural network (Cereghino et al., 2001). They were used to explain the observed structure of functional feeding groups of aquatic macro-invertebrates (Jorgensen et al., 2002); Self-Organizing Map (SOM) neural network was used to determine pest species assemblages for global regions (Worner and Gevrey, 2006); BP and RBF (radial basis function) neural networks were used to simulate and predict species richness of rice arthropods (Zhang and Barrion, 2006), reconstruct spatial pattern of insects (Zhang et al., 2007), and simulate survival dynamics of insects (Zhang and Zhang, 2008).

This study aimed to present several models, and evaluate their effectiveness in modeling spatial distribution of arthropods. Arthropods were investigated on the grassland. A neural network and a partial differential equation were developed to model above-ground distribution of arthropods. Models were validated and compared for their power in predictability. Details for developing and using neural network were discussed.

2 Materials and Methods

2.1. Field Investigation

Investigation was conducted on the grassland with 8×8 quadrates. Each quadrate has an area of 1×1 m^2. Arthropods were collected, identified, and counted for every quadrate. Insects were sorted and identified to order level and the other arthropods were identified to classis level.

2.2. Artificial Neural Network

The artificial neural network for modeling spatial distribution of arthropods is a mapping from input space (with the spatial coordinates of quadrate as the element) to output space (with number of arthropod individuals in the quadrate as the element), $U:R^2 \rightarrow R$ and $u(\boldsymbol{x})=v$, where $u \in U=\{u|u:R^2 \rightarrow R\}$. For an input set, $\boldsymbol{x}_i \in R^2$, and the output set, $v_i \in R$, there is a mapping f that satisfies $f(\boldsymbol{x}_i)=v_i$, $i=1,2,\ldots,n$. A mapping $u \in U=\{u|u:R^2 \rightarrow R\}$, represented by this network, should approximate $f(\boldsymbol{x})$ and satisfy the following condition:

$$|u(\boldsymbol{x})- f(\boldsymbol{x})| <\varepsilon \quad \boldsymbol{x} \in R^2$$

where $\boldsymbol{x}=(x, y)^T$, and$\varepsilon>0$ is the known threshold for error.

A three-layer neural network was developed for modeling spatial distribution of arthropods (Fig. 1). Both the first and second layers contained thirty neurons, and bias was used to each layer. Transfer functions for layers 1 tc 3 were hyperbolic tangent sigmoid transfer function:

$$\text{tansig}(\boldsymbol{x}) = 2/(1+\exp(-2*\boldsymbol{x}))-1,$$

logarithmic sigmoid transfer function:

$$\text{logsig}(\boldsymbol{x}) =1/(1+\exp(-\boldsymbol{x})),$$

and linear transfer function:

$$\text{purelin}(\boldsymbol{x}) = \boldsymbol{x},$$

respectively. Initialization of network, and weights and bias for each layer, was performed by a function that initializes each layer i (i=1,2,3) according to its own initialization function (Hagan et al., 1996; Mathworks, 2002; Fecit, 2003). Network was trained using Levenberg-Marquardt backpropagation algorithm. Desired performance function was mean squared error performance function (mse). The first and second layers received inputs from input space and produced outputs for the third layer. There was a closed loop for the third layer. For each layer, the net input functions calculated the layer's net input by combining its weighted inputs and biases.

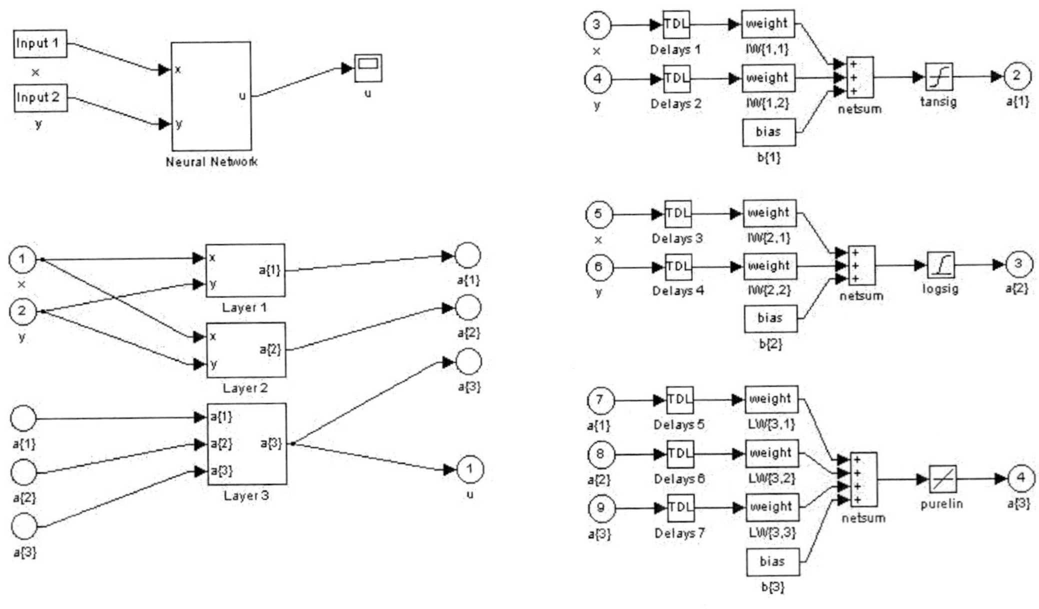

Figure 1. Artificial neural network developed in present study.

Mathematically, the network output is:

$$f(\boldsymbol{x}) \approx u(\boldsymbol{x}) = \sum_{k=1}^{3} \omega_k \, a_k(\cdot) \tag{1}$$

where

$$a_1(\cdot) = 2/(1+\exp(-2(\omega_{11}\, x + \omega_{21}\, y + b_1))) - 1$$

$$a_2(\cdot) = 1/(1+\exp(-(\omega_{12}\, x + \omega_{22}\, y + b_2)))$$

$$a_3(\cdot) = \sum_{k=1}^{3} \omega_{k3}\, a_k(\cdot) + b_3$$

In eq. (1), $\boldsymbol{x}=(x, y)^{T}$ is the input, $u=u(\boldsymbol{x})$ is the output; ω_i, i=1,2,3; ω_{ij}, i,j=1,2; ω_{i3}, i=1,2,3…, and b_i, i=1,2,3, are the parameters.

The artificial neural network was developed using Matlab (Mathworks, 2002). Modeling performance of the neural network was expressed as mse, Pearson correlation coefficient, and significance level for the linear regression between the simulated and observed. The Matlab algorithm for the neural network developed and used in this study are as follows:

```
clc
clear net;
data=arth;
n=30;    % Number of neurons
rows=8;
cols=8;
s=0;
for i=1:rows;
for j=1:cols;
s=s+1;
da(s,1)=j;da(s,2)=i;da(s,3)=data(i,j);
end;
end;
datax=da(:,1);
datay=da(:,2);
net=network;
net.numInputs=2;
net.numLayers=3;
net.inputs{1}.size=1;
net.inputs{2}.size=1;
net.biasConnect=[1;1;1];
net.inputConnect=[1 1;1 1;0 0];
net.layerConnect=[0 0 0;0 0 0;1 1 1];
net.outputConnect=[0 0 1];
net.targetConnect=[0 0 1];
mi=min(datax);ma=max(datax);
tt(1,1)=mi;tt(1,2)=ma;
net.inputs{1}.range=tt;
mi=min(datay);ma=max(datay);
tt(1,1)=mi;tt(1,2)=ma;
net.inputs{2}.range=tt;
net.layers{1}.size=n; net.layers{2}.size=n;
% Transfer functions: logsig,tansig,purelin,radbas,satlins,tribas
net.layers{1}.transferFcn='tansig';
net.layers{2}.transferFcn='logsig';
net.layers{1}.initFcn='initlay';
net.layers{2}.initFcn='initlay';net.layers{3}.initFcn='initlay';
net.inputWeights{2,1}.delays=[0 1];
net.inputWeights{1,2}.delays=[0 1];
net.layerWeights{3,3}.delays=1;
net.initFcn='initlay';
net.performFcn='mse';
net.trainFcn='trainlm';
```

```
net.trainParam.goal=1e-05;
net.trainParam.epochs=10000;
net=train(net,[da(:,1)';da(:,2)'],da(:,3)');
y=sim(net,[da(:,1)';da(:,2)']);
y
```

2.3. Response Surface Model (RSM)

The response surface model (He, 2001; Mathworks, 2002), i.e., the trend surface model (Zhang and Fang, 1982), was also used in present modeling:

$$u(\boldsymbol{x})=\text{a}+\mathbf{b}^T\boldsymbol{x}+\boldsymbol{x}^T\mathbf{c}\boldsymbol{x} \quad (2)$$

where $u(\boldsymbol{x})$: arthropod abundance (individuals per quadrate); $\boldsymbol{x}=(x, y)^T$: the coordinate of quadrate; $\mathbf{b}=(b_1,b_2)^T$, $\mathbf{c}=(c_1,c_2)^T$: parametric vectors; a: constant.

2.4. Spline Function

Spline interpolation is one of the most efficient interpolation models, among which the cubic spline interpolation is widely used (Burden and Faires, 2001). The cubic spline function used in present study was:

$$\begin{aligned} u(x)=&M_{i+1}(x-x_i)^3/(6l_i)+M_i(x_{i+1}-x)^3/(6l_i)+(f(x_{i+1})/l_i - M_{i+1}l_i/6)(x-x_i)+\\ &(f(x_i)/l_i - M_il_i/6)(x_{i+1}-x),\\ &x\in[x_i,x_{i+1}],\ i=0,1,\ldots,n-1 \end{aligned} \quad (3)$$

where, $x_i=i+1$, $M_i=S''(x_i)$, $i=0,1,\ldots,n$; $l_i=x_{i+1}-x_i$, $i=0,1,\ldots,n-1$. M_i, $i=0,1,\ldots,n$, were obtained from three-bending moment equation (Zhang, 2007b).

2.5. Partial Differential Equation

The following partial differential equation (PDE) was developed to model spatial distribution of arthropods in a limited region G with a boundary Γ:

$$\partial^2u/\partial x^2+p(x,y)\partial^2u/\partial y^2+q(x,y)\partial u/\partial x+v(x,y)\partial u/\partial y+w(x,y)u=f(x,y),\ (x,y)\in G$$

$$u|_\Gamma=\Phi(x,y),\ (x,y)\in\Gamma \quad (4)$$

where $u(x,y)$: the number of arthropod individuals in (x,y). $p(x,y)$, $q(x,y)$, $v(x,y)$, $w(x,y)$, and $f(x,y)$ are continuous functions on $G+\Gamma$.

Let $x_i=x_0+ih$, $y_j=y_0+j\tau$, $u_{i,j}=u(x_i,y_j)$, $p_{i,j}=p(x_i,y_j)$, $q_{i,j}=q(x_i,y_j)$, $v_{i,j}=v(x_i,y_j)$, $w_{i,j}=w(x_i, y_j)$, $f_{i,j}=f(x_i,y_j)$, $i=0,\pm1,\pm2,\ldots$; $j=0,\pm1,\pm2,\ldots$; then all differences were derived as the following:

$$(\partial^2u/\partial x^2)_{i,j}\approx(u_{i+1,j}-2u_{i,j}+u_{i-1,j})/h^2$$

$$(\partial^2 u/\partial y^2)_{i,j} \approx (u_{i,j+1} - 2u_{i,j} + u_{i,j-1})/\tau^2$$

$$(\partial u/\partial x)_{i,j} \approx (u_{i+1,j} - u_{i-1,j})/(2h)$$

$$(\partial u/\partial y)_{i,j} \approx (u_{i,j+1} - u_{i,j-1})/(2\tau)$$

and the difference equation for eq. (4) was expressed as the following:

$$(u_{i+1,j} - 2u_{i,j} + u_{i-1,j})/h^2 + p_{i,j}(u_{i,j+1} - 2u_{i,j} + u_{i,j-1})/\tau^2 + q_{i,j}(u_{i+1,j} - u_{i-1,j})/(2h) + \\ + v_{i,j}(u_{i,j+1} - u_{i,j-1})/(2\tau) + w_{i,j}u_{i,j} = f_{i,j} \quad (5)$$

In present study, $h=\tau=1$, and the functions $p(x,y)$, $q(x,y)$, $v(x,y)$, $w(x,y)$, and $f(x,y)$, were expressed as the following linear functions:

$$p(x,y) = a_p + b_p x + c_p y$$

$$q(x,y) = a_q + b_q x + c_q y$$

$$v(x,y) = a_v + b_v x + c_v y$$

$$w(x,y) = a_w + b_w x + c_w y$$

$$f(x,y) = a_f + b_f x + c_f y \quad (6)$$

The fifteen parameters in eq. (6) were obtained by fitting spatial distribution data with a group of linear equations based on eq. (5) and eq. (6).

2.6. Data Description

(1) Training data. In the modeling of spatial distribution, in total of 64 quadrates (n=64) were used to train neural network and response surface model. The input space was a two-dimensional space (coordinates of quadrate, e.g., (1,2), (5,7), etc.), and the output space was a one-dimensional space (arthropod abundance).

(2) Cross validation. There are several cross validation methods. I used a widely applicable method (Olden et al., 2006). Using this method, each quadrate was separately removed from the input set of 64 quadrates, and the remaining quadrates were used to train model and to predict the removed quadrates using the trained model. As a consequence, the cross validation may be conducted within the data set in the same study. Comparisons between the predicted and observed arthropod abundances were made and Pearson correlation coefficient (r) and statistic significance were calculated to validate models.

(3) Quadrates were submitted to neural network in two ways, i.e., their fixed sequences (Fig. 2), and randomized sequences of quadrates.

1	2	3	4	5	6	7	8
9	10	11	12	13	14	15	16
17	18	19	20	21	22	23	24
25	26	27	28	29	30	31	32
33	34	35	36	37	38	39	40
41	42	43	44	45	46	47	48
49	50	51	52	53	54	55	56
57	58	59	60	61	62	63	64

Figure 2. Fixed sequences of quadrates defined in present study.

3. Results

Most arthropods found on the grassland were insects, which belong to the orders Homoptera (523 individuals), Orthoptera (230 individuals), Hymenoptera (110 individuals), Coleoptera (55 individuals), and Diptera (40 individuals), etc. Other arthropods were sparsely distributed on the grassland. The spatial distribution of arthropods exhibited a saddle-like shape, which was similar to the most abundant Homoptera (Fig. 3).

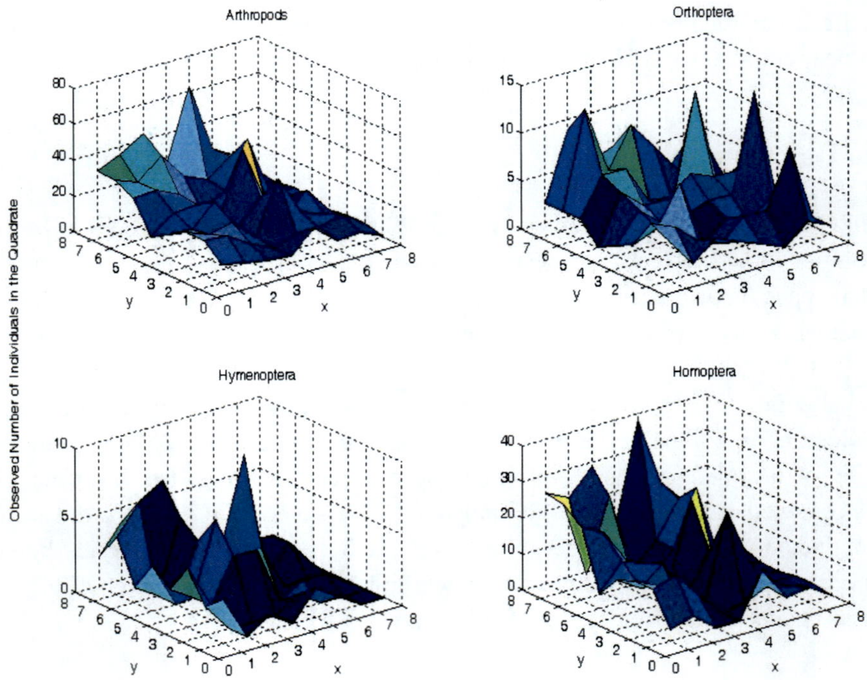

Figure 3. Observed spatial distribution of individuals of arthropods, Orthoptera, Hymenoptera, and Homoptera on the grassland.

3.1. Modeling Spatial Distribution with Neural Network and Response Surface Model

Using the artificial neural network developed above (Fig.1, eq.(1)) to simulate spatial distributions of arthropods and the most abundant orders Orthoptera, Hymenoptera, and Homoptera. Neural network was trained by 10000 epochs and the desired accuracy (mse) was 0.00001. The results revealed that neural network exhibited excellent simulation performance. The simulated spatial distribution perfectly coincided with the observed (intercept≈0, slope≈1, r≈1, p<0.0001), as illustrated by Fig. 4.

Using a deviation function, stmse=mse/u^2, where u is the averaged individuals per quadrate, together with Fig. 4, it was found that the lower abundance would overall lead neural network to yield the better simulation performance (Arthropods (stmse=$6.39*10^{-3}$), Homoptera (stmse=$2.16*10^{-2}$), Orthoptera (stmse=$5.88*10^{-7}$), Hymenoptera (stmse=$1.46*10^{-6}$)).

Response surface model could better fit the spatial distribution of arthropods.

However, its simulation performance was lower than neural network (Fig. 5).

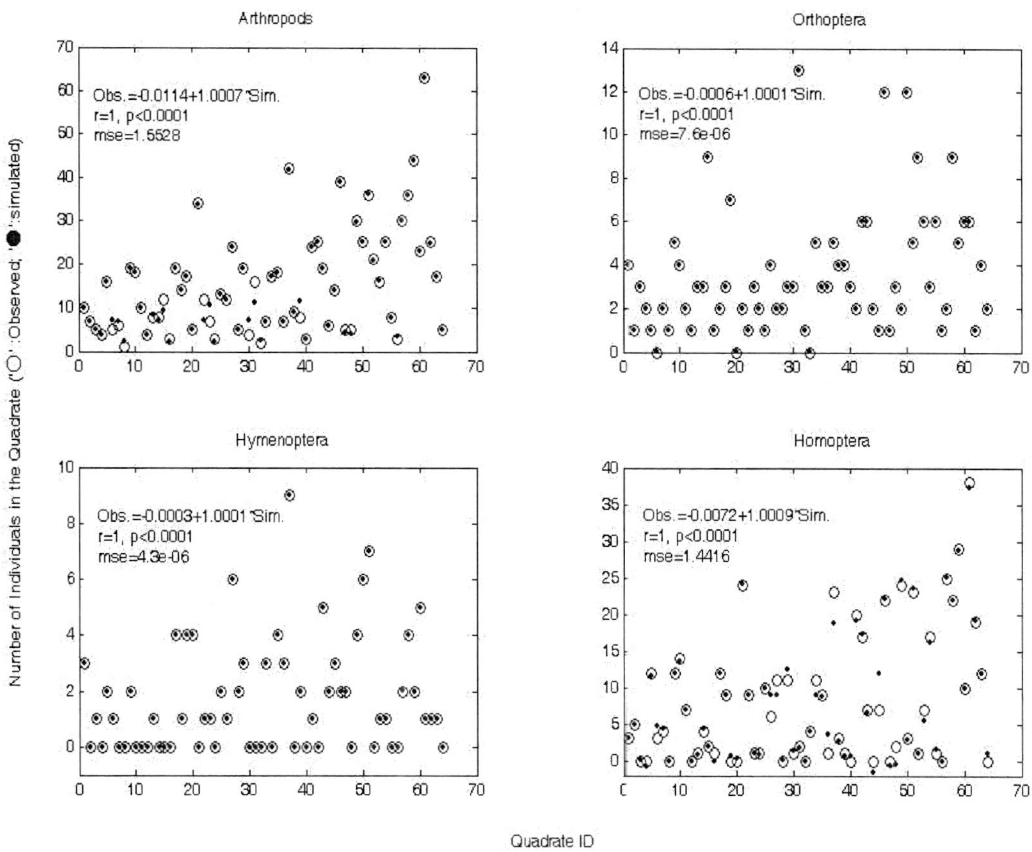

Figure 4. Simulating spatial distribution of arthropods using neural network. Quadrates were submitted to neural network in their fixed sequences.

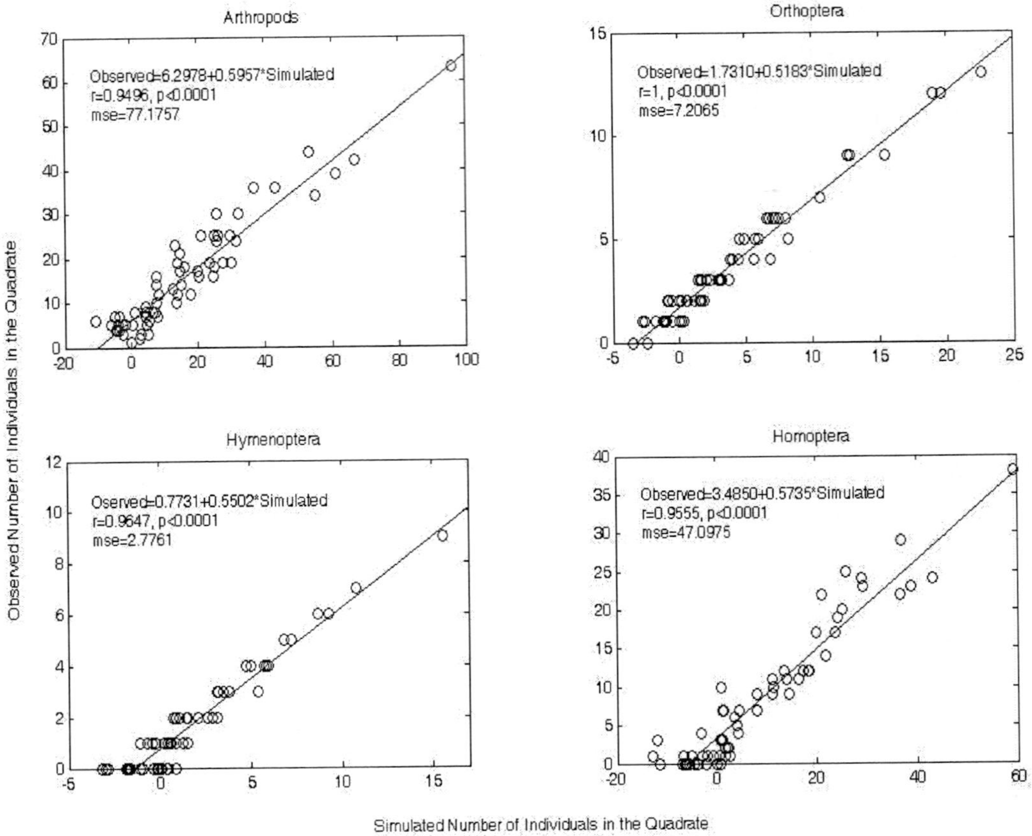

Figure 5. Simulating spatial distribution of arthropods using response surface model.

3.2. Cross Validation of Models

The cross validation, in which quadrates were submitted to neural network in their fixed sequences, demonstrated that neural network exhibited much better performance than response surface model in predicting unknown quadrates (Fig. 6). In most cases, response surface model yielded the negative correlation between the predicted and observed abundances. As a result, it is not suggested using response surface model to predict spatial distribution of arthropods.

Different from the situation above, neural network exhibited better prediction performance in the case of larger abundance than lower abundance on the grassland (Fig. 6). This means that compared to simulation the neural network needs more information to train itself for producing a reasonable extrapolation.

In an additional cross validation of neural network to predict the spatial distribution of arthropods, quadrates were submitted in randomized sequences and five randomizations were used. The results showed that neural network yielded the better performance (r=0.5323, p<0.0001). More than 60% of quadrates were correctly predicted and they fell inside 95% confidence intervals of the predicted (Fig. 7).

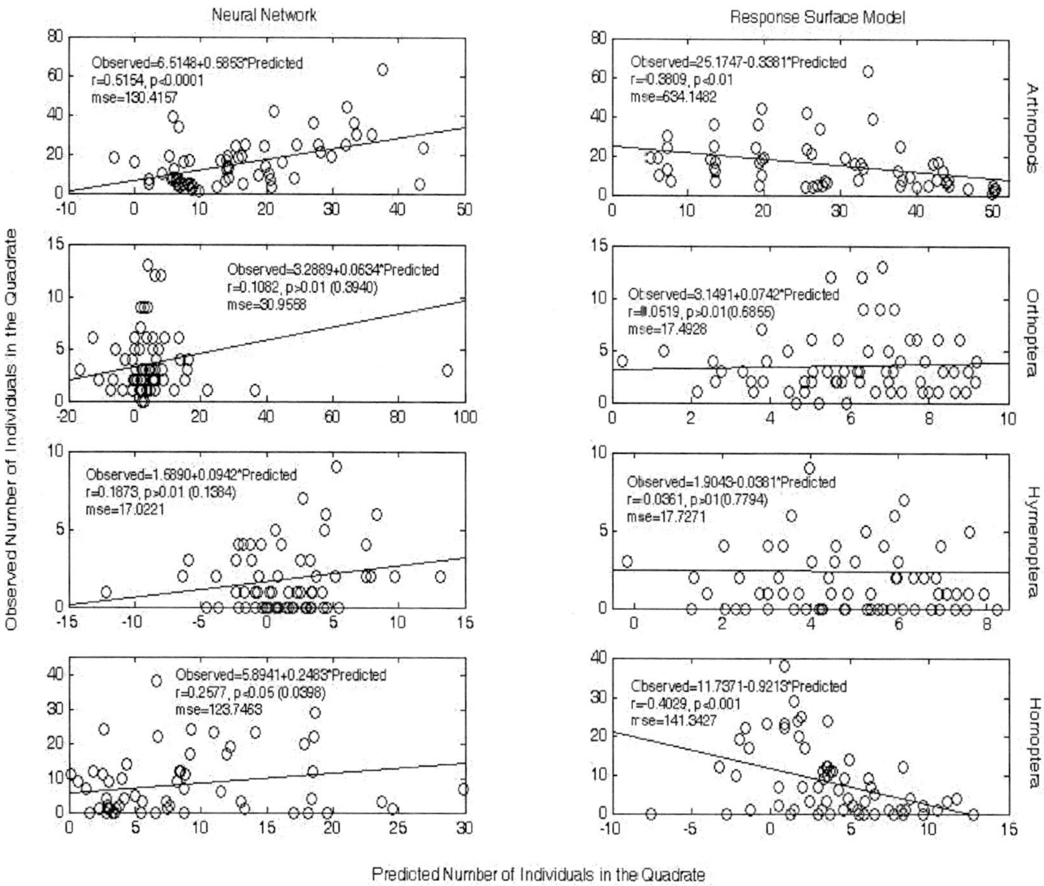

Figure 6. Cross validation of neural network and response surface model for predicting spatial distribution of arthropods. Quadrates were submitted to neural network in their fixed sequences.

The cross validation of spline function indicated that spline function performed badly than both neural network and response surface model (Table 1).

Table 1. Cross validation of spline interpolation

Arthropods	Orthoptera
Observed=15.6520-0.0041*Simulated *r*=-0.0100, *p*>0.01(0.9341), mse=1164.4	Observed=3.8052-0.0545 *r*=-0.0975, *p*>0.01(0.4436), mse=37.7524
Hymenoptera	**Homoptera**
Observed=1.7092+0.0067*Simulated *r*=0.0100, *p*>0.01(0.9289), mse=15.4195	Observed=8.3154-0.0148*Simulated *r*=-0.0316, *p*>0.01(0.8072), mse=445.9147

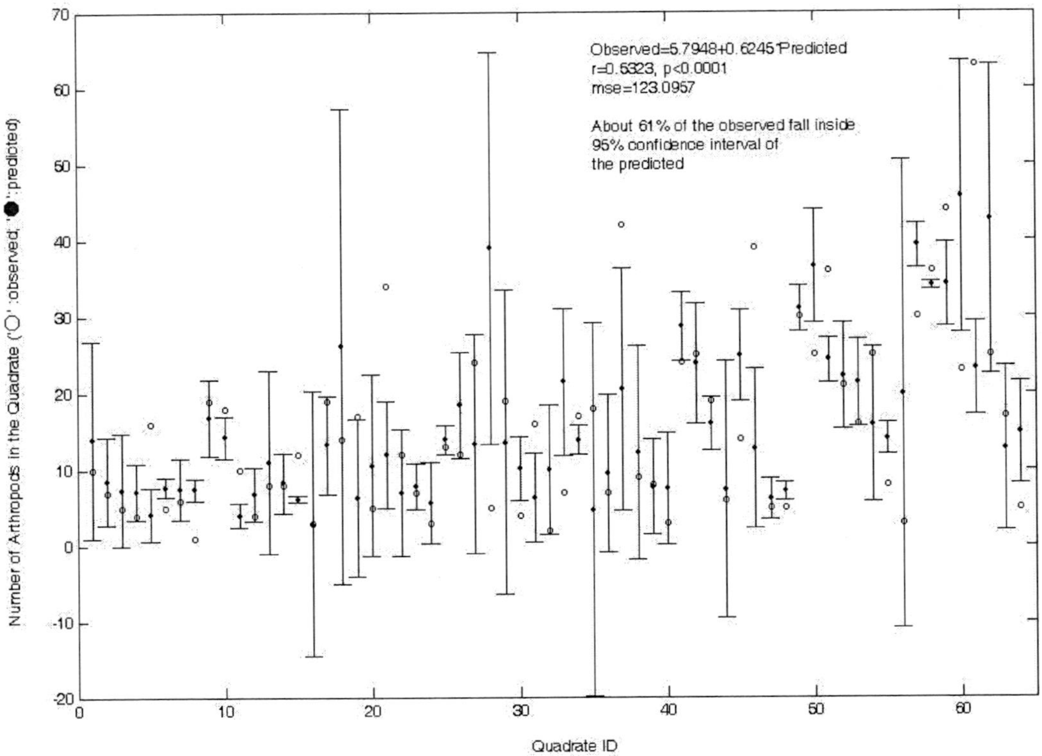

Figure 7. Cross validation of neural network for predicting distribution of arthropods. Quadrates were submitted to neural network in randomized sequences. Five randomizations were conducted.

3.3. Describing Spatial Distribution with PDE

A group of linear equations were developed according to eq. (5) and eq. (6), with calculated coefficient matrix from distribution data of arthropods. For total arthropods, and the most abundant three orders, Orthoptera, Hymenoptera, and Homoptera, their coefficient matrices (15×36) were fully ranked (rank=15). The parameters were obtained as indicated in Table 2, from which the partial differential equation can be obtained for specific taxonomic group. For example, the partial differential equation for spatial distribution of arthropods is to be the following:

$$\partial^2 u/\partial x^2+(1.254-0.065x-0.110y)\partial^2 u/\partial y^2+(-0.777+0.467x-0.547y)\partial u/\partial x+$$
$$+(2.848-0.555x-0.193y)\partial u/\partial y+(3.852-0.349x+0.111y)u=$$
$$=21.908+5.263x-0.551y,\ (x,y)\in G$$
$$u|_\Gamma=\Phi(x,y),\ (x,y)\in\Gamma$$

This equation may be used to describe the spatial distribution of arthropods. It fully fitted the spatial distribution. Theoretically this equation can be used to extrapolate spatial distribution of arthropods.

Table 2. Parameters in partial differential equation (eqs. (4)~(6))

	Arthropods	Orthoptera	Hymenoptera	Homoptera
a_p	1.2535	-0.5867	1.2676	1.6496
b_p	-0.0645	-0.0301	-0.0113	-0.2635
c_p	-0.1102	0.1902	-0.2396	-0.0236
a_q	-0.7771	-4.322	-0.8041	-1.6311
b_q	0.4671	0.4873	0.1044	0.216
c_q	-0.5469	0.2556	0.0068	-0.0558
a_v	2.8476	-0.7527	-0.0643	0.9427
b_v	-0.5548	-0.1756	0.0505	-0.3421
c_v	-0.1927	0.3112	-0.1443	0.0872
a_w	3.8523	-0.8764	3.0785	6.1561
b_w	-0.3495	0.0595	0.0512	-0.7088
c_w	0.1113	0.6596	-0.3247	-0.1529
a_f	21.9082	-10.6057	6.0289	32.9485
b_f	5.2626	1.3989	0.7485	-0.9004
c_f	-0.551	2.972	-1.1149	-2.8578

4. Conclusions and Discussion

An artificial neural network is developed to model spatial distribution. It exhibits excellent performance in the simulation of spatial distribution of arthropods. Simulated spatial distribution would perfectly coincide with the observed one. Lower total abundance would overall lead neural network to yield the better simulation performance. Response surface model and spline function can be used to fit the spatial distribution of arthropods. Simulation performance of response surface model is proved to be lower than neural network. The cross validation confirms that neural network has much better performance than response surface model and spline function in predicting unknown quadrates. Submitting quadrates in randomized sequences helps to yield confidence interval of the result in the neural network modeling. A partial differential equation is also developed to model spatial distribution. An algorithm for obtaining parametrical values in the equation is presented. The partial differential equation can be used to extrapolate spatial distribution of arthropods.

There are many kinds of artificial neural networks. More and more new algorithms of neural networks have been developing in various sciences. Both classic algorithms, like BP, SOM, RBF (Nagendra and Khare, 2006; Worner and Gevrey, 2006; Zhang and Barrion, 2006), etc., and newly developed algorithms such as the one developed in present study, can be used in the modeling of spatial distribution of arthropods. New algorithms may be designed according to the specific requirements and some details should be deliberated, such as the number of layers, neurons, biases, and targets; transfer functions, training functions; layer connects, input connects, output connects, bias connects, and target connects; input delays, input weights, layer weights, and initiate functions, and so on.

In addition to network settings, the quality of input set should also be ensured to obtain a better neural network. Data quality may be improved through a reasonable experiment or sampling design, eliminating data redundancy, and randomization procedure, etc (Kilic et al., 2007). As demonstrated in present study, randomized submission of quadrates helps neural network eliminate the correlation between inputs in training procedure. A larger number of randomizations are suggested being used to produce the reliable interval for simulation and prediction.

Over-learning in neural network modeling should be avoided in order to produce the best predictive performance. It can be solved by limiting the complexity of the neural network (layers, neurons, etc.), by training neural network with noise, and by using techniques as weight decay, and so on (Ozesmi et al., 2006).

Both spatial information (Euclidean coordinates) and environmental factors like plant composition, climate conditions, etc., can be considered to be input components. By doing this, we obtain a more interpretative neural network. This model may be used to interpret data. Some methods on data interpretation of explore neural network, like sensitivity analysis, inference rule extraction, randomization approach, neural interpretation diagram, etc., are available now for this purpose (Bradshaw et al., 2002; Olden and Jackson, 2002; Gevrey et al., 2006).

Similar to the conclusion from present study, most previous studies showed that neural networks outperfomed conventional models (Lek et al., 1997; Paruelo and Tomasel, 1997; Brosse et al., 1999; Abrahart and White, 2001; Yu et al., 2006; Zhang, 2007a,b), although varied results were also produced in using neural networks (Marchant and Onyango, 2003; Filippi and Jensen, 2006). This study demonstrates that artificial neural network is more robust than the conventional model in the modelling of spatial distribution. It can provide a feasible alternative to more classical spatial statistical techniques (Pearson et al., 2002). However, researches towards complicate spatial distribution are further desired in the future.

This study used the data set of 64 quadrates to model spatial distribution of arthropods and a better performance was achieved. As fitting models, their modeling performance is expected to be improved with the increase of the size of data set. However, more studies with larger data sets are still needed in the future to further validate these models.

Acknowledgments

This project was supported by “National Basic Research Program of China”(973 Program)(No. 2006CB102005), and “948” project of China (2006-G32). We thank all participants of arthropod investigation, Mr. WG Zhou, HQ Dai, and undergraduates of ecology 2004, Sun Yat-sen University, China.

References

Abdel-Aal RE. 2004.Hourly temperature forecasting using abductive networks. *Engineering Applications of Artificial Intelligence*, **17**:543-556

Abrahart RJ, White SM. 2001.Modelling Sediment Transfer in Malawi: Comparing Backpropagation Neural Network Solutions Against a Multiple Linear Regression Benchmark Using Small Data Set. *Phys. Chem. Earth (B)*, **26**(1): 19-24

Acharya C, Mohanty S, Sukla LB, et al. 2006. Prediction of sulphur removal with *Acidithiobacillus* sp. using artificial neural networks. *Ecological Modelling*, **190**(1-2): 223-230

Almasri MN, Kaluarachchi JJ. 2005. Modular neural networks to predict the nitrate distribution in ground water using the on-ground nitrogen loading and recharge data. *Environmental Modelling and Software*, **20**: 851-871

Ballester EB, Valls GCI, Carrasco-Rodriguez, JL, et al. 2002. Effective 1-day ahead prediction of hourly surface ozone concentrations in eastern Spain using linear models and neural networks. *Ecological Modelling*, **156** (1): 27-41

Bradshaw CJA, Davis LS, Purvis M, et al. 2002. Using artificial neural networks to model the suitability of coastline for breeding by New Zealand fur seals (Arctocephalus forsteri). *Ecological Modelling*, **148** (2): 111-131

Brown KS Jr. 1991. Conservation of Neotropical insects: Insects as indicators. In: Collins, NM, Thomas, JA (eds.), *The Conservation of Insects and Their Habitats*. Academic Press, London, pp.349-404

Burden R, Faires JD. *Numerical Analysis (*Seventh Edition). Thomson Learning, Inc., 2001

Cereghino R, Giraudel JL, Compin A. 2001. Spatial analysis of stream invertebrates distribution in the Adour-Garonne drainage basin (France), using Kohonen self organizing maps. *Ecological Modelling* ,**146**:167–180

Fecit. *Analysis and Design of Neural Networks in MATLAB 6.5,* Electronics Industry Press, Beijing, China, 2003

Filippi AM, Jensen JR. 2006. Fuzzy learning vector quantization for hyperspectral coastal vegetation classification. *Remote Sensing of Environment*, **100**:512-530

Gevrey M, Dimopoulos I, Lek S. 2006. Two-way interaction of input variables in the sensitivity analysis of neural network models. *Ecological Modelling*, **195** (1-2): 43-50

Hagan MT, Demuth HB, Beale MH. *Neural Network Design*. PWS Publishing Company, USA, 1996

He RB. MATLAB 6: *Engineering Computation and Applications*. Chongqing University Press, Chongqing, China, 2001

JØrgensen SE, Verdonschot P, Lek S. 2002. Explanation of the observed structure of functional feeding groups of aquatic macro-invertebrates by an ecological model and the maximum exergy principle. *Ecological Modelling*, **158** (3): 223-231

Kilic H, Soyupak S, Tuzun I et al. 2007. An automata networks based preprocessing technique for artificial neural network modelling of primary production levels in reservoirs. *Ecological Modelling*, **201** (3-4): 359-368

Kremen C, Colwell RK, Erwin TL, et al. 1993. Invertebrate assemblges: their use as indicators in conservation planning. *Conservation Biology*, **7**: 796-808.

Kuo JT, Hsieh MH, Lung WS et al. 2007. Using artificial neural network for reservoir eutrophication prediction. *Ecological Modelling*, **200**(1-2): 171-177

Lek S, Baran P. 1997. Estimations of trout density and biomass: a neural networks approach. *Nonlinear Analysis, Theory, Methods & Applications*, **30**(8): 4985-4990

Loot G, Giraudel JL, Lek S. 2002. A non-destructive morphometric technique to predict Ligula intestinalis L. plerocercoid load in roach (Rutilus rutilus L.) abdominal cavity. *Ecological Modelling*, **156** (1): 1-11

Marchant JA, Onyango CM. 2003. Comparison of a Bayesian classifier with a multilayer feed-forward neural network using the example of plant/weed/soil discrimination. Computers and Electronics in Agriculture, 39:3-22

Mathworks. 2002. *Neural Network Toolbox*, MATLAB 6.5

McKenna JE. 2005. Application of neural networks to prediction of fish diversity and salmonid production in the Lake Ontario basin. *Transactions of The American Fisheries Society*, **134**(1):28-43

Moisen GG, Frescino TS. 2002. Comparing five modelling techniques for predicting forest characteristics. *Ecological Modelling*, **157** (2-3): 209-225

Nagendra SMS, Khare M. 2006.Artificial neural network approach for modelling nitrogen dioxide dispersion from vehicular exhaust emissions. *Ecological Modelling*, **190**(1-2): 99-115

Nour MH, Smith DW, El-Din MG, et al. 2006. The application of artificial neural networks to flow and phosphorus dynamics in small streams on the Boreal Plain, with emphasis on the role of wetlands. *Ecological Modelling*, **191**(1): 19-32

Olden JD, Jackson DA. 2002. Illuminating the "black box": a randomization approach for understanding variable contributions in artificial neural networks. *Ecological Modelling*, **154** (1-2): 135-150

Olden JD, Joy MK, Death RG. 2006. Rediscovering the species in community-wide predictive modeling. *Ecological Applications*, **16** (4): 1449-1460

Ozesmi SL, Tan CO, Ozesmi U. 2006. Methodological issues in building, training, and testing artificial neural networks in ecological applications. *Ecological Modelling*,**195** (1-2): 83-93

Pastor-Barcenas O, Soria-Olivas E, Martın-Guerrero JD.2005. Unbiased sensitivity analysis and pruning techniques in neural networks for surface ozone modeling. *Ecological Modelling*, **182**:149–158

Pearson RG, Dawson TP, Berry PM, et al. 2002. SPECIES: A Spatial Evaluation of Climate Impact on the Envelope of Species. *Ecological Modelling*, **154** (3): 289-300

Pimental D, Stachow U, Takacs DA, et al. 1992. Conserving biological diversity in agricultural/forestry systems. *Bioscience*, **42**(5): 354-362

Schultz A, Wieland R. 1997. The use of neural networks in agroecologica modeling. *Computers and Electronics in Agriculture*, **18**:73-90

Sharma V, Negi SC, Rudra RP, et al. 2003. Neural networks for predicting nitrate-nitrogen in drainage water. *Agricultural Water Management*, **63**:169–183

Tan CO, Ozesmi U, Beklioglu M, et al. 2006. Predictive models in ecology: Comparison of performances and assessment of applicability. *Ecological Informatics*, **1**(2): 195-211

Viotti P, Liuti G, Di Genova P. 2002. Atmospheric urban pollution: applications of an artificial neural network (ANN) to the city of Perugia. *Ecological Modelling*, **148**(1): 27-46

Wilson EO. 1987. The little things that run the world. *Conservation Biology*, **1**:344-346

Worner SP, Gevrey M. 2006. Modelling global insect pest species assemblages to determine risk of invasion. *Journal of Applied Ecology*, **43** (5): 858-867

Yu R, Leung PS, Bienfang P. 2006. Predicting shrimp growth: Artificial neural network versus nonlinear regression models. *Aquacultural Engineering*, **34**:26–32

Zhang WJ. 2007a. Supervised neural network recognition of habitat zones of rice invertebrates. *Stochastic Environmental Research and Risk Assessment*, **21**: 729-735

Zhang WJ. 2007b. *Methodology on Ecology Research*. Sun Yat-sen University Press, Guangzhou, China

Zhang WJ, Barrion AT. 2006. Function approximation and documentation of sampling data using artificial neural networks. *Environmental Monitoring and Assessment*, **122**: 185-201

Zhang WJ, Bai CJ, Liu GD. 2007. Neural network modeling of ecosystems: a case study on cabbage growth system. *Ecological Modelling*, **201**:317-325

Zhang WJ, Liu GH, Dai HQ. 2008. Simulation of food intake dynamics of holometabolous insect using functional link artificial neural network. *Stochastic Environmental Research and Risk Assessment*, **22**: 123-133

Zhang WJ, Zhang XY. 2008. Neural network modeling of survival dynamics of holometabolous insects: a case study. *Ecological Modelling*, **211**: 433-443

Zhang WJ, Zhong XQ, Liu GH. 2008. Recognizing spatial distribution patterns of grassland insects: neural network approaches. *Stochastic Environmental Research and Risk Assessment*, **22**: 207-216

Zhang YT, Fang KT. 1982. *Introduction to Multivariate Statistics*. Science Press, Beijing, China, 1982

In: Cluster Computing and Multi-Hop Network Research ISBN 978-1-60876-186-9
Editors: C. Jimenez and M. Ortego, pp. 73-90

Chapter 4

Parallel Molecular Dynamics Simulation using a Hybrid Force Field with Quantum Chemical Charge Calculations

Olaf Schneider**, *Ulrich Sternberg and Alexander Vondrous
Karlsruhe Institute of Technology (KIT), Forschungszentrum Karlsruhe GmbH,
P. O. Box 36 40, 76021 Karlsruhe, Germany.

Abstract

The simulation of biomolecules like proteins in their natural environment is computational demanding and parallelization is a must to simulate relevant time scales within a reasonable computation time. The software COSMOS uses a mixed approach of classical molecular dynamics (MD) and quantum chemistry (QC) by incorporating atomic charges computed using the Bond Polarization Theory (BPT) into the force field. To parallelize the MD we used an atom-decomposition approach with replicated data. A kind of greedy algorithm yields a partition of the atoms without dividing chemical bonds. The parallel version of the force field COSMOS-NMR is implemented using MPI. We present the results of performance measurements. As test case we used an ice crystal heated over a picosecond. The computations are performed on a HPC cluster with InfiniBand. It turns out, that our code greatly benefits from the low latency of the interconnect.

1. Introduction

The computer-aided determination and simulation of molecular structures is a key issue for comprehensive understanding of biochemical processes. Such an understanding is worthwhile, for example, in drug design to understand the therapeutic mechanism. Typically, not only the target molecule itself, such as a protein, needs to be simulated but also its interaction with the environment: cell membranes, cytosol or other aqueous fluids. This results in a complex molecular system with a huge number of atoms whose dynamic behavior is worth studying. Other objectives are molecular properties such as chemical shifts or atomic charges.

*E-mail address: olaf.schneider@kit.edu

Standard simulation techniques range from "ab initio" approaches, which fully exploit the knowledge of quantum chemistry (QC), to classical molecular dynamics (MD), which uses Newton's mechanics within a parameterized force field. "ab initio" in this context refers to the first principles, upon the calculations are based, meaning a certain level of approximation that is used in standard quantum chemical programs for molecular computations. By determining the electronic structure of the molecules QC methods gain a maximum of predictive power, whereas classical MD yields reasonable results for many problems at much lower computational costs. Intermediate are QC methods based on approximations and empirical data known as semi-empirical and empirical methods. Density functional theory (DFT) and simulation methods based on it are gaining more and more popularity in recent. Most authors subsume DFT under ab-initio methods, even it uses an empirical exchange potential. The vast majority of MD codes uses pair potentials to describe the forces between atoms. Force field parametrization is usually done via experimental data or ab-initio calculations. Hybrid QM/MM methods try to combine advantages of quantum-mechanical (QM) and molecular-mechanical (MM) methods [21], that is, to achieve the accuracy of quantum chemistry models with the computational speed of MD by treating regions of interest with QM methods and the rest of the system with a MM force field (see [7, 21]).

The force field COSMOS-NMR [15, 30] combines MM methods with QC calculations. Specifically it uses the so called Bond Polarization Theory (BPT) [29] which was first introduced to calculate NMR chemical shifts. To obtain a description for the electron density distribution in BPT calculations for every bond a bond orbital is introduced. Therefore no explicit QM region is needed and the whole system can be treated in the same manner. This method proofed to be applicable to the partial atomic charges that are obtained as the solution of a linear system of equations [32]. The BPT establishes a linear dependence of the charges from the bond polarization energy. Using sets of empirical bond parameters atomic properties like partial charges can be calculated from bond polarization energies. The parameters are obtained from ab initio calculations. This approach is especially beneficial if NMR data like chemical shifts are of interest. For example, it is possible to adjust the 3D structure of a molecule to NMR measurements [39].

As another unique feature, the COSMOS-NMR force field is applicable to wide range of substance classes. The calculation of non bonding interactions of condensed systems can be performed with periodic boundary conditions. That is, one or, if necessary, two shells of boxes surrounding the central simulation box can be taken into account. In solids these boxes are taken as the crystalline unit cells.

The software COSMOS [31] was originally created for MS Windows using C++. A graphical user interface (GUI) allows to view and edit molecular models, to move and rotate them, to optimize their geometry and to calculate various other molecular properties in a simple manner. Later on, a command-line-oriented backend without GUI was developed in order to use high performance computers for extensive calculations. This backend is ported onto various UNIX platforms.[1] To make use of clusters and massively parallel computers in an efficient manner we had to parallelize the compute intensive parts of the code.

[1]The COSMOS backend program is supplied free for scientific applications by one of the authors (requests to U. Sternberg, `postmaster@COSMOS-software.de`).

There exists parallel implementations of MD for several state-of-the-art force fields, such as CHARMM [2, 3, 13], NAMD [17, 18], LAMMPS [12, 20] as well as GROMACS [10], EULERGROMOS [5], and other descendants of the force field GROMOS (see also [6]). Other researchers have reported on novel MD implementations for various kinds of parallel computers (see [14, 19]).

Many MD implementations uses the Verlet [35] velocity time integration method to solve Newton's equations of motion. The force vector is computed as the gradient of a sum over pair potentials between atoms. Typically, the non-bonded interactions of atoms such as Coulomb forces are most computational demanding. To reduce the computational costs one introduces cutoff distances for the forces, often combined with special damping functions to avoid discontinuities in the force field. Another commonality is the use of pair lists to organize the calculation.

A parallel implementation of MD requires the parallelization of the additive pair potentials. Regarding the literature on proposed and implemented algorithms, we can distinguish two orthogonal decisions in the design: first, *atom-decomposition* versus *force-decomposition*, and second, *distributed data* versus *replicated data*. In the first strategy, the atoms are partitioned and each processor only performs those calculations needed to compute the motion of its assigned atoms. The alternative relies on a force-decomposition by partitioning the list of pairwise interacting atoms. If this partitioned list is distributed over the processors of a distributed-memory computer, one easily achieves initial load-balancing (cf. [6]). During the simulation, lists needs to be updated if atoms approach each other and fall below the cutoff radius. Additional programming efforts are required to avoid load imbalance. Distributed data means that each processor in a cluster or a distributed-memory parallel computer has direct access only its assigned atoms or forces. If a quantity from another processor is needed during calculations, it is fetched by message passing. Replicated data, on the other hand, stores all atomic data on all processors. An intermediate approach is the spatial decomposition of the atomic system into small 3D boxes, one attached to each processor (cf. [5, 19, 33]). To perform force calculations, atomic coordinates are stored replicated at the boundary. Thus, the overlap size relates to the maximal range of force coverage. It is not unusual, that the overlap covers the complete neighboring box, hence resulting in a 3D grid with mesh size chosen somewhat smaller then the cutoff distance.

Plimpton [20] describes a version of parallel MD, which is very similar that one realized for COSMOS-NMR. However, the considered force field does not incorporate charge calculations. The main feature of this approach are replicated stored atomic coordinates on each compute node. Thus, each processor has access to the coordinates and properties of all atoms without additional communication overhead. Vincent and Merz [36] reported on a version of AMBER4 based on force-decomposition together with replicated data. They attributed the replicated data strategy for parallel MD to Smith and his coworkers (see [25–27]).

Unlike the MD codes mentioned so far, COSMOS-NMR incorporates elaborated calculations of the charge distribution into the force field (cf. Section 2.5.). Consequently, the computational amount of a single time integration step is significantly higher than in other MD implementations. The benefit is a much more accurate approximation of the Coulomb forces resulting in more reliable simulations since all mutual polarizations are included into the electrostatic energy.

The remainder of this chapter is organized as follows: Section 2. and its various subsections present the parallel variant of COSMOS-NMR in details, starting with an note on embarrassingly parallel calculations. Our basic algorithm for parallel MD is presented in Section 2.2., followed by the algorithm to establish the partition of the molecular system in Section 2.3.. Thereafter, communication overhead of MD (Section 2.4.) and specifics of the charge calculation (Section 2.5.) are discussed. Eventually, we present the results of our performance analysis in Section 3.. The paper is completed with conclusions, final remarks and references to future work in Section 4..

2. Parallel COSMOS-NMR

2.1. Embarrassingly Parallel Computations

Even without the newly implemented message passing based parallelization, the COSMOS backend may benefit from running in a compute cluster with several nodes. Apart from MD simulations the backend supports several other calculation modes. Among them is an optimization of molecular geometry by minimizing the potential energy of the molecular system. An optimization method may exploit the available derivatives of the potential. Thus, a conjugate gradient method with a sophisticated adaption of step-size scale factors is applied to the potential energy surface. Real-world examples usually demand energy minimization for a whole bunch atomic coordinate sets. Therefore the backend possesses an operation protocol to perform multiple geometry optimizations at once. Actually, this task was the original motivation to separate the calculations from the graphical user interface. Obviously, several geometry optimizations can run in parallel if a sufficient number of processors is available. The implementation is straight forward, no communication between the tasks is needed.

2.2. Parallel MD

As in any other MD code, potential energy calculations within the force field, specifically the non-bonded interactions of atoms, are are most computational demanding. Thus, this part of code requires attention. When parallelizing for distributed-memory computers, layout and distribution of data structures is the crucial point.

The basic idea was replicated storing of atomic coordinates and other properties. Each parallel task is assigned to a disjoint subset of atoms and computes only the forces affecting this atoms. Using atom-decomposition and replicated data was already suggested by other researchers (cf. [11, 20]), however, distinct specifics of COSMOS-NMR guided us to this solution.

Unlike the codes referred in the introduction, COSMOS-NMR does not form explicit pair lists to carry out the energy and force calculations. Instead, for every single atom we compute the impact of all other atoms in the atomic system. This saves the administration overhead for the pair lists. Additionally, it provides us with greater flexibility in the code. For example, we can select a subset of atoms and simulate their motion, while all other atomic coordinates remains fixed. Nevertheless, the impact of the fixed atoms on the moved ones is accumulated into the force field. In this way, the parallel MD tasks of COSMOS-

```
setup p parallel tasks for MD calculation
replicate atomic coordinates and properties
generate a partition of the atoms corresponding to the p tasks
do p parallel :
    ateach timestep do
        update atomic charges (cf. Section 2.5.)
        foreach selected atom on this task do
            compute potential energy, forces, velocities,
            and new atomic coordinates
            send the computed atomic coordinates to other tasks
        end
        receive new atomic coordinates for all other atoms
    end
end
```

Algorithm 1: Pseudo-code of parallel MD with the COSMOS-NMR force field.

NMR was implemented with only minimal modifications to the serial code. Moreover, once operations on subsets of atoms are implemented this can be performed for periodic boundary conditions or for fixing bond distances with SHAKE [22] and all these calculations will run in a distributed parallel fashion (see also [27]). The downside of this simplicity is, that enhancements, like the cutoff radius for short-distance forces, reduces only the number of arithmetical operations but not the overall loop count. Moreover, many terms in the force field are computed twice. In the serial version, this can be avoided by adjusting loop counts (triangular run). When running in parallel, the duplicated force calculations are outweighed by the gained parallel performance. Additionally, using the molecular partition described in Section 2.3., we reduce the probability to find nearby atoms in different subsets.

The basic idea of our parallel MD is summarized in Algorithm 1.

2.3. Molecular Partition

The parallelization of COSMOS-NMR relies on a partition of the atomic system. To keep the implementation of the parallel MD as simple as possible, we require, that atoms connected with a chemical bond belong to the same parallel task. Thus, the atomic system is partitioned into subsets of molecules. In order to achieve a (nearly) load-balanced algorithm, the number of atoms in each cluster should be (nearly) equal. This *molecular partition* is constructed by a kind of greedy algorithm, which is outlined in Algorithm 2. Roughly speaking, at each step we assign the largest remaining molecule to the subset with the smallest number of atoms. The actual implementation uses heavily containers and algorithms of the C++ Standard Template Library (STL).

```
constitute a list of molecules,
wherein each molecule comprises a list of atoms
sort molecules by decreasing number of atoms
initialize p empty lists of molecules (one for each task)
foreach selected atom on this task do
    select the partial list with the smallest number of atoms
    put the current atom into this list
end
```

Algorithm 2: Greedy algorithm for the molecular partition.

Beyond this heuristic approach we can model this task as a standard set partitioning problem from operations research:

$$\begin{aligned} &\text{minimize } \boldsymbol{c}^\top \boldsymbol{x} && \text{over } \boldsymbol{x} = \left[x_j\right]_{j=1}^{n}, && x_j \in \{0,1\} \\ &\text{subject to } B\boldsymbol{x} = \boldsymbol{1}, && \text{where } B = \left[b_{i,j}\right]_{i=1,j=1}^{m,n}, && b_{i,j} \in \{0,1\} \end{aligned}$$

where $\boldsymbol{x}$ is an incidence vector which characterizes the solution of the problem, $\boldsymbol{c}$ is a cost vector, B is an incidence matrix whose columns describes the feasible subsets and $\boldsymbol{1}$ is a column vector of suitable dimension with one in each component (cf. [4]).

Our objective is to partition M molecules into p disjoint subsets. Each molecule is equipped with an integer w_i, the number of atoms in the molecule. We denote by N the total number of atoms in the system, i. e. $N = \sum_{i=1}^{M} w_i$. Consider an arbitrary subset $\mathcal{F} \subseteq \{1, \ldots, M\}$ of molecules. It is characterized by an incidence vector $\boldsymbol{f} = \left[f_1, \ldots f_M\right]$, consisting of zeros and ones, where $f_i = 1$ indicates that the ith molecule belongs to the subset $\mathcal{F}$. In our setting, all subsets of molecules are feasible, thus the matrix B contains a column $\boldsymbol{f}$ for each subset $\mathcal{F}$. The equality in the constrain ensure that each molecule is allocated to exactly one subset.

The cost vector should reflect our request of an balanced partition in the sense that the total number of atoms per subset should vary as little as possible. The optimal load balancing is achieved with N/p atoms per subset, which may be impossible for a given set of molecules. However, the deviation of the actual number of atoms in a subset from the ideal value N/p is a measure of the costs:

$$c(B) = \left| \frac{N}{p} - \sum_{i=1}^{M} f_i w_i \right| = \left| \sum_{i=1}^{M} f_i w_i - \frac{1}{p} \sum_{i=1}^{M} w_i \right| .$$

Thus we have defined all quantities of the set partitioning problem and may apply any standard method from operations research to solve it.

2.4. Communication Overhead of the Replicated Data Scheme

The purpose of this section is to provide an estimation of the communication overhead induced by the parallelization scheme described above. This theoretical estimation will be compared to our measurement later on. Since the partition described in Section 2.3. is only performed once in the start-up phase of the program, it is neglected in this considerations.

Of course, the main question is, how much communication overhead increases if more processors are used in the calculation and compare this with the reduction of calculation time gained. As already mentioned, the computational amount of MD is dominated by the calculation of non-bonded interactions between the atoms in the force field. For n atoms the costs are proportional to n^2, which is expressed as $O(n^2)$ in Landau's Big Oh notation. Practically, with COSMOS-NMR we observed for 39 atoms a calculation time of less than 100 milliseconds per step and about 30 seconds with 663 atoms (cf. Section 3.2.).

By virtue of the replicated data approach, each processor has direct access to all atomic coordinates and properties. Communication is only needed to exchange the newly computed atomic coordinates in each timestep. To be precise, a second communication step with similar overhead is involved in the charge calculation (cf. Section 2.5.).

We denote by N the total number of atoms within the system and the calculation is performed on p parallel tasks. Further, we assume that the partitions of molecules is ideally balanced, such that every tasks has $n = N/p$ dedicated atoms (which implies that N is an integer multiple of p).

If the message exchange works as indicated in Algorithm 1, lines 9 and 11, each task has to perform $p - 1$ send-receive pairs. The size of each message is proportional to n, the number of atoms per task. Thus, we estimate $O((p-1)n) = O(\frac{p-1}{p}N) \lessapprox O(N)$ as communication bandwidth overhead at each single task. Since all tasks communicate at the same time, the latency depends on the total number of communicating pairs of tasks, i. e. $O(p(p-1)/2)$.

Taking all together, the computational and communication costs behaves as:

$$O(\tfrac{N^2}{p^2}) + O(\tfrac{p-1}{p}N) + O(\tfrac{p(p-1)}{2}).$$

At first glance, the number of atoms should be greater than the squared number of processors to ensure that communication costs does not majorize calculation. In practice however, the first term is equipped with a constant factor in the order of milliseconds, while the other terms range in the order of microseconds. Present-day cluster networks like InfiniBand have a bandwidth of nearly a gigabyte per second and the point-to-point latency at MPI level go down to about 4 μs (cf. [24]). Since message size in our MD code is far below a kilobyte per atom, the second cost term above can be neglected.

An alternative approach to exchange the atomic coordinates would be to pick one parallel task as master, which collects the coordinates sent by all other tasks and redistributes them afterwards. The needed bandwidth at the master task is the same as above. The slaves send one coordinate data set of size $n = N/p$ and receive $p - 1$, resulting in $O(N)$ again. Latency is a little bit different: since only one process has to communicate with $p-1$ others, we get $O(p - 1)$. Although it seems the second approach is better, it should be noted, that the master serial performs all send and receive operations while the slaves idle. Whereas the first approach allows to overlap send and receive times on the different tasks.

On the other hand, it can be done even better: Plimpton [20] describes a communication pattern based on a binary tree resulting in $O(\log_2(P))$ sends and receives per processor with an accumulated message size of $O(N - N/p)$. In either case, there is a theoretical limit of scalability for the replicated data scheme, since the communication overhead for the global exchange of data grows with the number of processors.

However, in our code the exchange of atomic coordinates is implemented by an opaque MPI [16] subroutine executing an all-to-all communication. Thus, the actual communication overhead depend on the internals of the MPI library. However, our estimation presented in this section should be interpreted as worst case scenarios. Any serious MPI implementation will do better.

2.5. Charge Calculation

The week point of traditional force fields is the calculation of interaction energies with fixed atomic charges. If molecules approach each other to form complexes their mutual polarization will enhance the interaction energy. Therefore many attempts have been made to introduce this effect into the treatment of the electrostatics (for a short review see Tan et al. [34] and the references therein). All these methods, like electronegativity equlization, induced dipoles, or Drude oscillators, have one feature in common: additional empirical formulae are introduced into the force fields including the need of parametrization.

A unique feature of the COSMOS-NMR force field is the adaptive calculation of atomic charges based on bond polarization theory (BPT) [29, 32]. The charges calculated via BPT depend on the three-dimensional structure of the molecule. Since this structure varies over subsequent MD timesteps, the atomic charges have to be recalculated. The charges are calculated by solving a system of linear equations, which is deduced as follows:

$$\begin{aligned} q_{\mathrm{A}} &= \langle \psi | \hat{q}_{\mathrm{A}} | \psi \rangle \\ &= \sum_{\mathrm{A} \stackrel{i}{\longrightarrow} \mathrm{B}} \left(n_i p_i + n_i^2 A_i \sum_{\substack{\mathrm{X} \in C \\ \mathrm{X} \neq \mathrm{A}}} \left\{ \left\langle \chi_{\mathrm{A}}^i \middle| \frac{q_{\mathrm{X}}}{|\vec{R}_{\mathrm{X}} - \vec{r}|} \middle| \chi_{\mathrm{A}}^i \right\rangle - \left\langle \chi_{\mathrm{B}}^i \middle| \frac{q_{\mathrm{X}}}{|\vec{R}_{\mathrm{X}} - \vec{r}|} \middle| \chi_{\mathrm{B}}^i \right\rangle \right\} \right) \end{aligned} \tag{1}$$

q_{A} denotes the atomic charge of atom A, which is according to BPT calculated as the expected value of the charge operator $\hat{q}_{\mathrm{A}}$ and the the wave function ψ of the considered atomic configuration C. The wave function ψ is build up as a configuration interaction development of Slater determinants of bond orbitals. These bond orbitals are linear combinations of hybrid atomic orbitals χ_{A}^i and χ_{B}^i. The relation $\mathrm{A} \stackrel{i}{\rightarrow} \mathrm{B}$ is used to indicate that the atoms A and B are bonded; the bond itself is denoted by i. n_i denotes the occupation number of the bond i and p_i is the bond partial charge. The parameter A_i quantifies the change of the charge due to bond polarization. Thus, in the outer sum we add up the contributions of each bond of A to the total charge, which consists of the bond partial charge corrected with the polarization of the bond caused by the other atoms. Therefore, the inner sum in the right-most term of (1) goes over all atoms in the configuration C expect A itself.

The parameters p_i and A_i are determined a-priori from "ab initio" calculated atomic charges for every combination of elements A and B and the two bond types σ and π. Provided a sufficient large set of charges is known the parameter set is the solution of an overdetermined system of equations. In our case the parametrization was performed on partial atomic charges from a set of 163 molecules that were optimized using an "ab initio" program [37, 38].

Regarding equation (1) it is obvious that q_{A} has to be calculated from all other atomic charges q_{X} of the molecular system. Since the parameters q_{X} can be taken as factors out of the integrals of the expected value, we arrive at a system of equations whose number of rows and columns is equal to the number of atoms:

$$\boldsymbol{F}\boldsymbol{x} = \boldsymbol{b}, \qquad \boldsymbol{x} = \left[q_{\mathrm{X}}\right]_{\mathrm{X}\in\mathcal{C}}, \qquad F = \left[F_{\mathrm{A,X}}\right]_{\mathrm{A,X}\in\mathcal{C}}, \qquad \boldsymbol{b} = \left[b_{\mathrm{A}}\right]_{\mathrm{A}\in\mathcal{C}} \tag{2}$$

where

$$F_{\mathrm{A,X}} = \begin{cases} 1 & \text{if A} = \text{X}, \\ -\sum_{\mathrm{A}\xrightarrow{i}\mathrm{B}} n_i^2 A_i (f_{\mathrm{A,X}} - f_{\mathrm{B,X}}) & \text{else,} \end{cases} \qquad b_{\mathrm{A}} = \sum_{\mathrm{A}\xrightarrow{i}\mathrm{B}} n_i p_i$$

$$f_{\mathrm{X,Y}} = \left\langle \chi_{\mathrm{Y}}^{i} \middle| \frac{1}{|\vec{R}_{\mathrm{X}} - \vec{r}|} \middle| \chi_{\mathrm{Y}}^{i} \right\rangle.$$

As a consequence of charge conservation, the total charge $\sum_{\mathrm{X}\in\mathcal{C}} q_{\mathrm{X}}$ of the system $\mathcal{C}$ has to remain constant in subsequent timesteps.

In order to reduce the computational costs associated with assembling and solving this linear system of equations, one can specify – already in the serial implementation – that the BPT charge calculation within COSMOS-NMR is performed only at every ℓth timestep. It could be argued, that charges change only slightly if the atoms move with moderate speed.

Within the current parallel implementation, the charge calculation is modified such that at each task we compute the polarization for the bonds of the selected atoms, while the charges of all other atoms are taken from the previous step and remain unaltered. Let $\mathcal{S}$ denote the set of selected atoms. Then we get for an atom $\mathrm{A} \in \mathcal{S}$:

$$q_{\mathrm{A}} = \sum_{\mathrm{A}\xrightarrow{i}\mathrm{B}} \left(n_i p_i + n_i^2 A_i q_{\mathrm{X}} \sum_{\mathrm{X}\in\mathcal{C}\setminus\mathcal{S}} (f_{\mathrm{A,X}} - f_{\mathrm{B,X}}) \right) + \dots$$
$$\dots + \sum_{\mathrm{A}\xrightarrow{i}\mathrm{B}} \left(n_i^2 A_i q_{\mathrm{X}} \sum_{\substack{\mathrm{X}\in\mathcal{S} \\ \mathrm{X}\neq\mathrm{A}}} (f_{\mathrm{A,X}} - f_{\mathrm{B,X}}) \right).$$

We end up with a smaller linear system having as much unknowns as there are selected atoms within $\mathcal{S}$:

$$\begin{aligned} F_{\mathcal{S}}\boldsymbol{x}_{\mathcal{S}} &= \boldsymbol{b}_{\mathcal{S}}, & \boldsymbol{x}_{\mathcal{S}} &= \left[q_{\mathrm{X}}\right]_{\mathrm{X}\in\mathcal{S}} \\ F_{\mathcal{S}} &= \left[F_{\mathrm{A,X}}\right]_{\mathrm{A,X}\in\mathcal{S}} & \boldsymbol{b}_{\mathcal{S}} &= \left[b_{\mathrm{A}}^{\mathcal{S}}\right]_{\mathrm{A}\in\mathcal{S}} \end{aligned}$$
$$b_{\mathrm{A}}^{\mathcal{S}} = \sum_{\mathrm{A}\xrightarrow{i}\mathrm{B}} \left(n_i p_i + n_i^2 A_i \sum_{\mathrm{X}\in\mathcal{C}\setminus\mathcal{S}} q_{\mathrm{X}} (f_{\mathrm{A,X}} - f_{\mathrm{B,X}}) . \right) \tag{3}$$

All remaining quantities are defined as above. Thus, step 6 of Algorithm 1 could be spelled out as: If the timestep counter is an integer multiple of ℓ, then assemble and solve the linear system (3), send the newly computed charges to the $p-1$ other tasks, receive new charges from $p-1$ other tasks replacing the old ones. The communication overhead is increased by another all-to-all communication.

The procedure described above can be interpreted as one step of the block Jacobi [9] iterative solver for the linear system (2) where the initial guess is the charge from the previous timestep. Arguing again, that the charges vary slowly from step to step, we can expect an adequate accuracy even with only one iteration step. The charge conservation can be used to measure the quality of this approximation.

Of course, it would be possible to solve the originally linear system directly. A naive attempt would be to do this at one task and use message passing to send the resulting charges to the others tasks. Though, assembling the total coefficient matrix is highly memory intensive and might be impossible on a single compute node. Therefore, a promising effort is to use a parallel solver for dense matrices, such as implemented in the ScaLAPACK [1] library.

3. Performance Analysis

3.1. Test Case

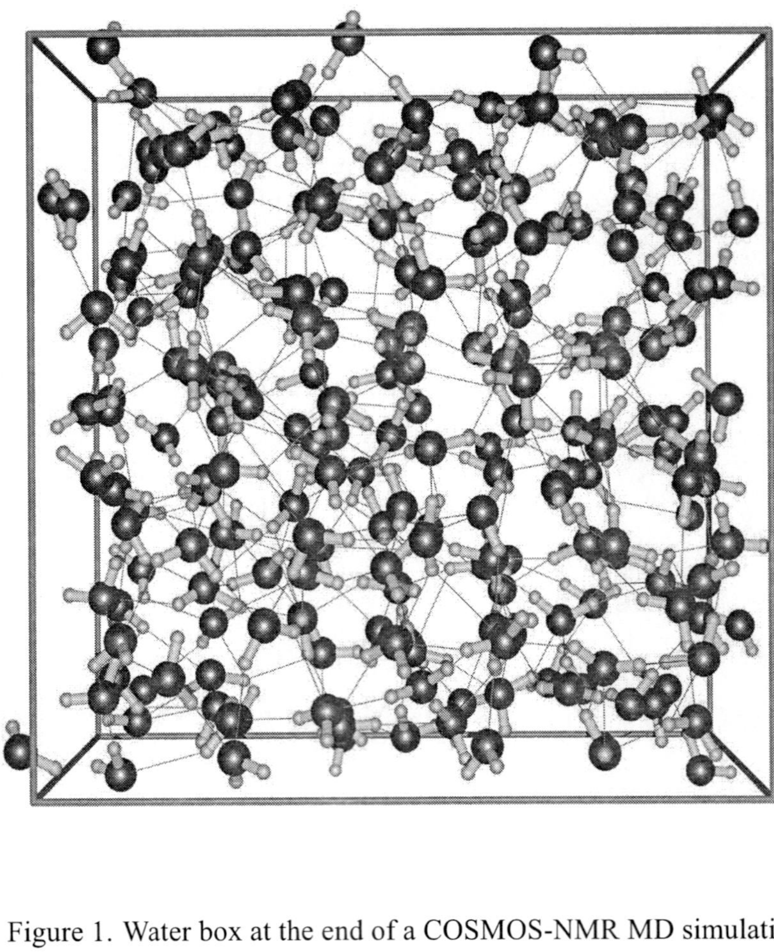

Figure 1. Water box at the end of a COSMOS-NMR MD simulation.

The task chosen for our performance measurement was to simulate melting of an ice configuration with 221 water molecules (see Fig. 1). The simulation has been done with a varying number of processors. 2000 timesteps are calculated within each simulation. Periodic boundary conditions were enforced and the cutoff distance for the Coulomb forces was similar to the box size, such that one layer of periodic boxes was sufficient. The atomic coordinates are transferred after each timestep with MPI via an InfiniBand connection. Each task is calculating the atomic charges in each timestep separately. After each 100th timestep the charges are exchanged between the parallel tasks. An intermediate result (coordinate and velocity snapshot) is saved after each 100th timestep. The positions of atoms within the molecule change on the time scale of the integration step and therefore the charges within molecules should be calculated in every timestep. Intermolecular interaction change on a much longer time scale and these smaller contributions can be therefore updated at a lower rate. By means of the molecular partition, this results in a very accurate provision for changes in charge due to polarization of bonds by atoms in the same molecule. Whereas, the less influential contributions of most other atoms are incorporated with a lower frequency to reduce communication costs.

The simulations on water are of high interest since water is the universal solvent of biologic systems. Here we will only refer to a recent work that studied the performance of QM/MM methods for the description of the properties of water [40]. In these QM/MM MD simulations one or two shells of water molecules that surround a central water molecule were embedded in a large MM box. The inner region is treated on several levels of QM approximation (see Table 1). Two shells of water in a QM are regarded as necessary to obtain a satisfactory description of the water properties. MP2 calculations give a much better description of intermolecular interactions but are too demanding to include two water spheres in a MD calculation. Since DFT includes electron correlation its results should be similar to MP2 but if larger water samples are investigated, the results become unreliable.

Table 1. Comparison of water parameters for box simulations using COSMOS-NMR with other QM/MM calculations; 1st row: first maxima of the $\mathrm{O}\cdots\mathrm{O}$ radial distribution functions (pair correlation functions), 2nd row: second maxima of the $\mathrm{O}\cdots\mathrm{H}$ radial distribution functions (cf. Fig. 2)

	$R_{\mathrm{O}\cdots\mathrm{O}}/\text{Å}$	$R_{\mathrm{O}\cdots\mathrm{H}}/\text{Å}$
COSMOS-NMR / BPT	2.93	1.93
QM/MM HF^a (two shells) [40]	2.92	2.06
QM/MM MP2^b (one shell) [40]	2.87	1.93
QM/MM B3LYP^c (one shell) [40]	2.81	1.85

^{a}Hatree-Fock level

^{b}Second order Møller-Plesset perturbation theory

^{c}DFT calculation using the B3LYP exchange potential

In the COSMOS-NMR simulation all 221 water molecules are treated on the same level of BPT approximation. Periodic box conditions were applied to the BPT atomic charges to ensure that all water molecules are treated are under the influence of the same potential. For the force field parameters of the polarizable and flexible water model we used the standard parameters for O H bonds and angles (see [15, 30]). The simulations are performed at 293 K for 1 ps with an time integration step size of 0.5 fs.

Comparing statistical parameters of the water box with other QM/MM calculations (Table 1) we obtain similar results. For an accurate description of hydrogen bonds the polarization of the water molecule and its electrostatics has to be treated correctly. Comparing the radial $O \cdots H$ distribution functions of our simulation (see Fig. 2) with those in [40] the results are highly similar. Especially, if in their calculation the QM region was extended to two water shells, their results match our calculations. The first peak around 1 Å(the top of the peak intensity was cut off in the figure) comes from the $O - H$ bond distances and the second peak from the $O \cdots H$ hydrogen bonds.

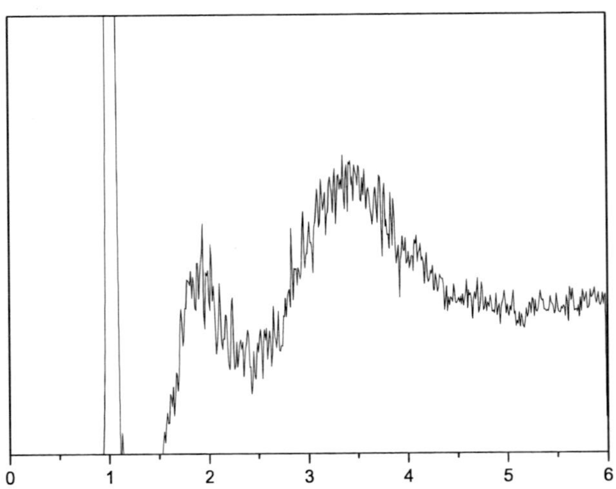

Figure 2. Radial distribution of $O \cdots H$ distances in a 1 ps water box simulation.

Partial radial distribution functions (as the $O \cdots H$ distribution shown in Figure 2) can be obtained from neutron diffraction experiments. Though, extracting the actual distribution function from measurements is not straightforward because of systematic uncertainties within the neutron diffraction data. For water a recent analysis was performed by Soper [28], who obtained $R_{O \cdots O} = 2.78$ Å and $R_{O \cdots H} = 1.79$ Å (cf. Table 1). Depending on the method of the diffraction data analysis distances are obtained that differ by about ± 0.1 Å from the values above (see [40]).

3.2. Measurements

Measured execution times are listed in Table 2 and depicted in Figure 3.

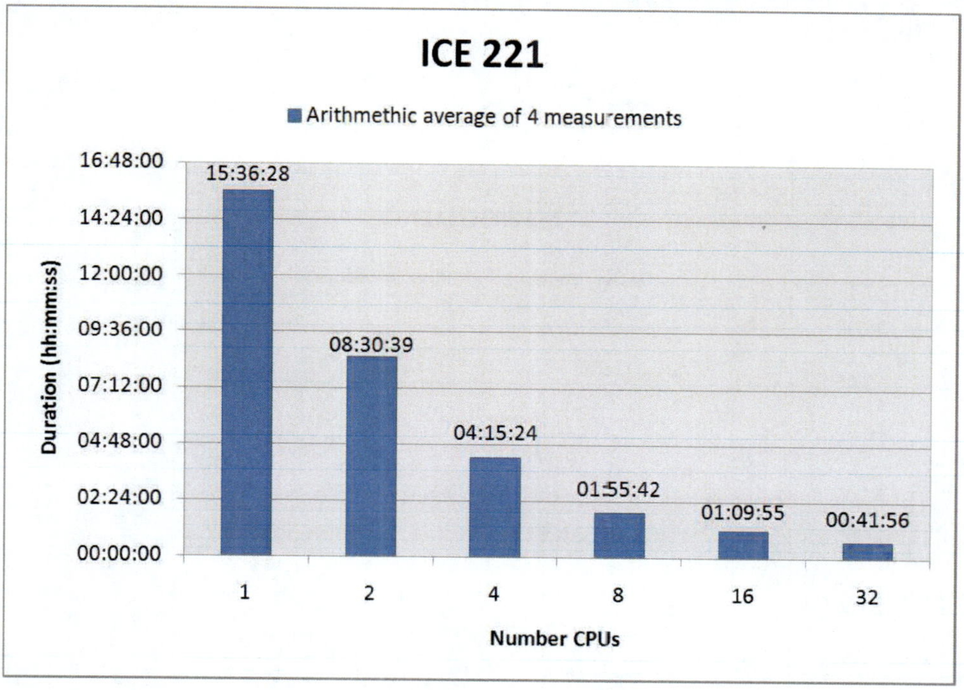

Figure 3. Execution times of parallel MD with varying number of processors.

Table 2. Execution times in hours:minutes:seconds from various runs

32	16	8	4	2	1
00:45:58	00:58:07	01:57:48	03:52:49	10:41:26	14:55:59
00:40:10	01:20:57	01:53:58	03:53:07	07:46:45	15:55:02
00:40:14	01:20:55	01:54:23	03:56:00	07:53:08	15:44:59
00:41:23	00:59:43	01:56:37	05:19:40	07:41:16	15:49:51

All timings are performed on the cluster Opus^{IB} [23]. The worker node hardware consisted of dual Opteron boards with 2.2 GHz processor clock cycle, 1 MB L1 cache and at least 2 GB of main memory. The InfiniBand interconnect is made up by 4 channels (4X SDR) with an theoretical overall transfer rate of 10 Gbit/s. Mentionable components of the software stack were Scientific Linux operating system, Portland Group compilers and the message passing library MVAPICH build upon a vendor-supplied InfiniBand driver stack.

The timing jobs were scheduled within the normal mixed operation of the cluster. Each task of our MPI job was assigned to a different node, but the second processor on a node was able to be used concurrently by another job.

Additional measurements are done with an smaller configuration consisting of only 13 molecules (39 atoms) where all other simulation parameters remained unchanged. Parting them on two tasks resulted in an total execution time of 06:16 minutes.

3.3. Discussion

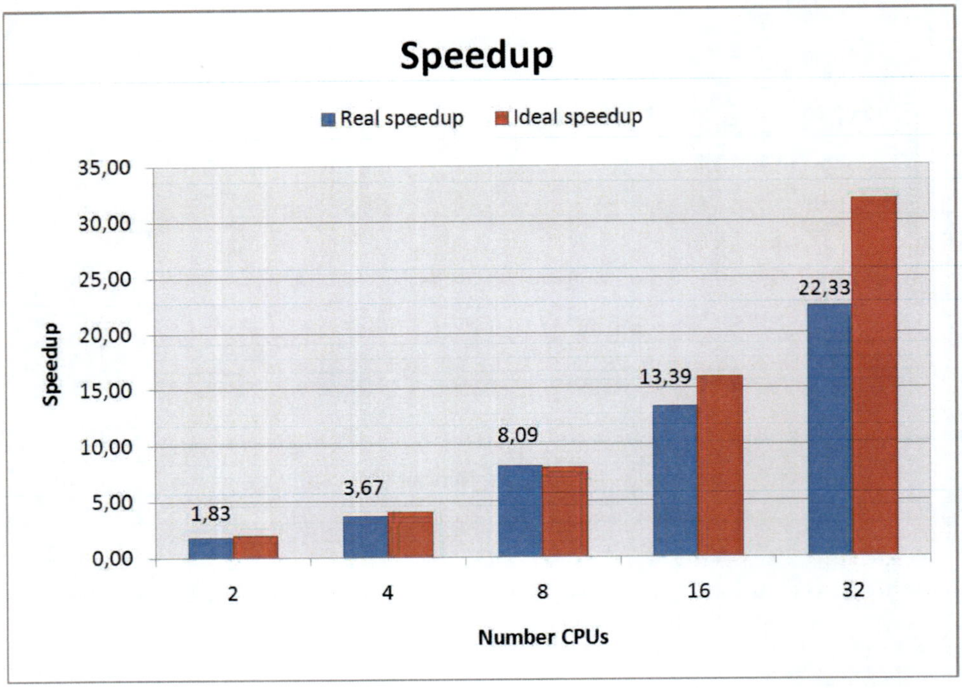

Figure 4. Parallel speedup of COSMOS MD.

Gained speedup is illustrated in Figure 4. A somewhat surprising point in this chart is the speedup of 8.09 with 8 CPUs. It can be explained by cache effects: If the number of used CPUs increases, the amount of memory per CPU is reduced. In the case of 8 CPUs, the amount of memory is that much reduced, that the CPU cache can be used much more efficient.

Even apart from those effects, the speedup seems better then it could be expected from our estimations in Section 2.4.. The communication overhead with an increasing number of tasks is outweigh by the saved computation time.

Note that our greedy algorithm supplied an optimal partition, for example with 32 CPUs we got 29 tasks with 21 atoms (7 molecules) and 3 tasks with 18 atoms (6 molecules). Each of this atoms interacts with all other atoms in the box plus and the additional boxes introduced by the periodic boundary conditions, but due to cutoff not all of these interacting forces needs to be computed. This computational amount is much higher then the time required for communication. Load balance is not ideal but well enough to not spoil the gained performance.

4. Conclusion

The presented parallelization of the COSMOS-NMR force field performs satisfying on the test case. Both, the calculation of non-bonded energies and forces in the MD as well as the BPT charge calculation are effectively parallelized. Thus, it was possible for the first time to parallelize a hybrid QM/MM molecular simulation. The molecular partition using the greedy algorithm achieves nearly equal load balance. For our test case we observed an good speedup, indicating that the theoretical limitations of the chosen parallelization scheme has no effect for the considered problem size. All these findings give rise to the hope, that we can simulate larger systems of atoms on longer time scales using the parallelized COSMOS program.

Though, our implementation could be enhanced in several ways: implementing an optimized communication pattern (binary tree) for the exchange of atomic coordinates, incorporating a parallel dense solver for the charge calculation, and parallel calculation of further NMR properties, such as chemical shifts.

Gridification is another direction of in the development of the software COSMOS, which we did not mention in this paper. The aim is to encapsulate the backend command-line calls into a Web Service, which is running on the head node of a cluster and can be invoked from within the GUI client on the desktop. A prototype of the service is already implemented using the Globus Toolkit [8].

References

[1] L. S. Blackford, J. Choi, A. Cleary, E. D'Azevedo, J. Demmel, I. Dhillon, J. Dongarra, S. Hammarling, G. Henry, A. Petitet, K. Stanley, D. Walker, and R. C. Whaley. *ScaLAPACK Users' Guide*. Society for Industrial and Applied Mathematics, Philadelphia, PA, 1997.

[2] B. Brooks, R. Bruccoleri, D. Olafson, D. States, S. Swaminathan, and M. Karplus. CHARMM: A program for macromolecular energy, minimization, and dynamics calculations. *Journal of Computational Chemistry*, **4**:187–217, 1983.

[3] CHARMM (Chemistry at HARvard Macromolecular Mechanics). `http://www.charmm.org/`.

[4] S. Chopra and D. Simchi-Levi. Packing and covering. In K. Rosen, editor, *Handbook of Discrete and Combinatorial Mathematics*, chapter 15.3, pages 996–1005. CRC Press, 2000.

[5] T. Clark, R. von Hanxleden, J. McCammon, and L. Scott. Parallelizing molecular dynamics using spatial decomposition. *Scalable High-Performance Computing Conference, 1994., Proceedings of the*, pages 95–102, May 1994.

[6] T. W. Clark and J. A. McCammon. Parallelization of a molecular dynamics nonbonded force algorithm for MIMD architecture. *Computers & Chemistry*, **14**(3):219 – 224, 1990.

[7] M. J. Field, P. A. Bash, and M. Karplus. A combined quantum mechanical and molecular mechanical potential for molecular dynamics simulations. *J. Comput. Chem.*, **11**(6):700–733, 1990.

[8] I. T. Foster. Globus Toolkit Version 4: Software for Service-Oriented Systems. In H. Jin, D. A. Reed, and W. Jiang, editors, *NPC*, volume 3779 of *Lecture Notes in Computer Science*, pages 2–13. Springer, 2005.

[9] G. Golub and C. F. Van Loan. *Matrix computations*. The Johns Hopkins Univ. Press, Baltimore, MD, 3rd edition, 1996.

[10] GROMACS: Fast, Free and Flexible MD. http://www.gromacs.de/.

[11] T. Karakasidis, N. Cholevas, and A. Liakopoulos. Parallel short range molecular dynamics simulations on computer clusters: Performance evaluation and modeling. *Mathematical and Computer Modelling*, **42**(7-8):783 – 798, 2005. International Conference of Computational Methods in Sciences and Engineering 2003.

[12] LAMMPS Molecular Dynamics Simulator. `http://lammps.sandia.gov/`.

[13] A. MacKerel Jr., C. Brooks III, L. Nilsson, B. Roux, Y. Won, and M. Karplus. *CHARMM: The Energy Function and Its Parameterization with an Overview of the Program*, volume 1 of *The Encyclopedia of Computational Chemistry*, pages 271–277. John Wiley & Sons: Chichester, 1998.

[14] J. Mederski, L. Mikulski, and P. Bala. Asynchronous parallel molecular dynamics simulations. In R. Wyrzykowski, J. Dongarra, K. Karczewski, and J. Wasniewski, editors, *Parallel Processing and Applied Mathematics*, volume 4967 of *Lecture Notes in Computer Science*. Springer Berlin / Heidelberg, 2008. 7th International Conference, PPAM 2007, Gdansk, Poland, September 9-12, 2007 Revised Selected Papers.

[15] M. Möllhoff and U. Sternberg. Molecular mechanics with fluctuating atomic charges - a new force field with a semi-empirical charge calculation. *Journal of Molecular Modeling*, **7**(4):90–102, 2001.

[16] MPI: Message Passing Interface. `http://www-unix.mcs.anl.gov/mpi/`.

[17] NAMS Scalable Molecular Dynamics. `http://www.ks.uiuc.edu/Research/namd/`.

[18] J. C. Phillips, R. Braun, W. Wang, J. Gumbart, E. Tajkhorshid, E. Villa, C. Chipot, R. D. Skeel, L. Kale, and K. Schulten. Scalable molecular dynamics with NAMD. *Journal of Computational Chemistry*, **26**:1781–1802, 2005.

[19] M. R. S. Pinches, D. J. Tildesley, and W. Smith. Large scale molecular dynamics on parallel computers using the link-cell algorithm. *Molecular Simulation*, **6**(1):51–87, 1991.

[20] S. J. Plimpton. Fast parallel algorithm for short-range molecular dynamics. *Journal of Computational Physics*, **117**:1–19, 1995.

[21] D. Riccardi, P. Schaefer, Yang, H. Yu, N. Ghosh, X. Prat-Resina, P. Konig, G. Li, D. Xu, H. Guo, M. Elstner, and Q. Cui. Development of effective quantum mechanical/molecular mechanical (qm/mm) methods for complex biological processes. *The Journal of Physical Chemistry B*, **110**(13):6458–6469, 2006.

[22] J.-P. Ryckaert, G. Ciccotti, and H. J. C. Berendsen. Numerical integration of the Cartesian equations of motion of a system with constraints: molecular dynamics of n-alkanes. *Journal of Computational Physics*, **23**(3):327–341, March 1977.

[23] O. Schneider, F. Schmitz, I. Kondov, and T. Brandel. Opus^{IB} – Grid enabled Opteron cluster with InfiniBand interconnect. *Lecture Notes in Computer Science*, **4699**:840–849, 2007.

[24] U. Schwickerath and A. Heiss. InfiniBand – Experiences at the Forschungszentrum Karlsruhe. *Nuclear Instruments and Methods in Physics Research Section A: Accelerators, Spectrometers, Detectors and Associated Equipment*, 559(1):81 – 84, 2006. Proceedings of the X International Workshop on Advanced Computing and Analysis Techniques in Physics Research - ACAT 05.

[25] W. Smith. Molecular dynamics on hypercube parallel computers. *Computer Physics Communications*, **62**(2-3):229 – 248, 1991.

[26] W. Smith. Molecular dynamics on distributed memory (MIMD) parallel computers. *Theoretica Chimica Acta*, **84**:385–398, 1993.

[27] W. Smith and T. R. Forester. Parallel macromolecular simulations and the replicated data strategy. *Computer Physics Communications*, **79**(1):52 – 77, 1994.

[28] A. K. Soper. The radial distribution functions of water and ice from 220 to 673 k and at pressures up to 400 mpa. *Chemical Physics*, **258**(2-3):121–137, August 2000.

[29] U. Sternberg. Theory of the influence of the second co-ordination sphere on the chemical shift. *Molecular Physics*, **63**(2):249–267, 1988.

[30] U. Sternberg, F.-T. Koch, M. Bräuer, M. Kunert, and E. Anders. Molecular mechanics for zinc complexes with fluctuating atomic charges. *Journal of Molecular Modeling*, **7**(4):54–64, 2001.

[31] U. Sternberg, F.-T. Koch, and P. Losso. COSMOS Pro 5: Computer Simulation of Molecular Structures, 2006. `http://www.cosmos-software.de/`.

[32] U. Sternberg, F.-T. K. Koch, and M. Möllhoff. New approach to the semiempirical calculation of atomic charges for polypeptides and large molecular systems. *Journal of Computational Chemistry*, **15**(5):524–531, 1994.

[33] T. P. Straatsma, M. Philippopoulos, and J. A. McCammon. Nwchem: Exploiting parallelism in molecular simulations. *Computer Physics Communications*, **128**(1-2):377 – 385, 2000.

[34] Y.-H. Tan, C. Tan, J. Wang, and R. Luo. Continuum polarizable force field within the poisson-boltzmann framework. *The Journal of Physical Chemistry. B*, **112**(25):7675–7688, June 2008.

[35] L. Verlet. Computer "Experiments" on Classical Fluids. II. Equilibrium Correlation Functions. *Phys. Rev.*, **165**(1):201–214, Jan 1968.

[36] J. J. Vincent and K. M. Merz Jr. A highly portable parallel implementation of AMBER4 using the message passing interface standard. *Journal of Computational Chemistry*, **16**(11):1420–1427, 1995.

[37] R. Witter. *Structure Elucidation with the COSMOS-NMR Force Field.* PhD thesis, Universität, Jena, 2003.

[38] R. Witter, M. Möllhoff, F.-T. Koch, and U. Sternberg. Fast atomic charge calculation to implement a polarisable force field - application to an ion channel protein. in preparation.

[39] R. Witter, W. Prieß, and U. Sternberg. Chemical shift driven molecular dynamics and geometry optimization. *Journal of Computational Chemistry*, **23**:298–305, 2002.

[40] D. Xenides, B. R. Randolf, and B. M. Rode. Structure and ultrafast dynamics of liquid water: A quantum mechanics/molecular mechanics molecular dynamics simulations study, *J Chem Phys*, **122**(17):174506, 2005.

Reviewed by:
Wolfgang Wenzel, Institut für Theoretische Physik, Universität Karlsruhe (TH)

In: Cluster Computing and Multi-Hop Network Research ISBN 978-1-60876-186-9
Editors: C. Jimenez and M. Ortego, pp. 91-108

Chapter 5

A New Clustering Algorithm for Mixed Data and Replicates

***I. Irigoien*[1] *and C. Arenas*[2*]**
[1]Department of Computation Sciences and Artificial Intelligence, University of Basque Country, Spain
[2]Department of Statistics, University of Barcelona, Spain

Abstract

In the biological and biomedical sciences, large multivariate datasets are frequently generated by investigators which include mixed data and replicate experiments. Recent technological advances, such as cDNA microarray technology have made possible the study of thousands of genes in a biological specimen. Given the flood of data that is generated from such experiments, appropriate analysis need to be performed so that useful knowledge and trends that would be otherwise undetectable can be observed. Our aim of interest is concerned with cluster analysis in a very general situation, in presence of mixed variables and replicates. To obtain several replicates is a straight forward method for improving the quality of inferences made from experimental studies. It is known that the replicates are not true duplicates, so the between-replicates variability must be taken in consideration. The usual practice is to average the replicate results before assessing differences between objects. This results in the loss of valuable information about variability. We describe a new partitioning cluster procedure based on a variant of Rao's diversity coefficient and in the so called path-distance. The approach uses top-down analysis. It starts with a large cluster and splits into smaller ones until each sub-cluster contains only one data point. The algorithm assigns each object to only one cluster, gives the global extreme (not local) of the function to be optimized and does not depend on the initial object selected to begin the algorithm with. The procedure takes into account the variance of the replicates. The methodology is a general algorithm that can be used to identify clusters in any situation, where data are arranged in tables (one for each replicate), where rows represent all objects for which data has been collected and columns represent the measured variables (binary, qualitative or continuous). We have tested the procedure on synthetic and on real data, obtaining good results and providing an easy-to-use bioinformatics solution for the clustering analysis of experiments with replicates. The program implementing the algorithm is available upon request from the authors.

*E-mail address: carenas@ub.edu. (Corresponding author.)

1. Introduction

The methodology presented in this paper has general applications but was motivated for the study of genetic and biomedical situations, where the experiments have been replicated or the data is not necessary continuous. In classification, one is concerned with assigning objects to classes on the basis of measurements made on these objects. In cluster analysis the classes are unknown *a priori* and need to be discovered from the data. An important application of cluster analysis is to cluster genes according to their expression profile which is obtained by cDNA microarray technology. A number of methods have been proposed for clustering objects as the agglomerative or divisive hierarchical methods [1], [2], or the partitioning clustering methods for example, k-means clustering [3], [4], [5]. Some clustering methods suppose a statistical model generally based on the multivariate normal distribution [6], [7], but this could limite their application. Other developed approaches can be found in the literature [8-25]. Recently, [26] developed some clustering algorithms based on Minimum Spanning Tree (MST), a concept from the graph theory, which have been implemented in the computer program EXCAVATOR [27], [28] which use iteration between MST based clustering and feature selection to remove noise. While all these approaches have clearly demonstrated their usefulness in applications [29], the majority of them are not appropriate for replicated experiments. In microarray analysis, two types of replication can be carried out: technical replication, when mRNA from a single biological case is used on multiple microarrays, and biological replication, when measurements are taken from multiple cases. Although early microarray experiments used few or no biological replicates, their necessity is now undisputed. Technical replicates allow only the effects of measurement variability to be estimated and reduced, whereas biological replicates allow this to be done for both measurement variability and biological differences between cases. There is some literature on the use of repeated measurements in clustering [30-34].

Another important question in real experiments is the type of data. As in biomedical applications [35] the data can be continuous measures (like color microarray experiments), measures of binary attributes (tumor presence or absence) and measures of multi-state attributes (like the classical blood groups) or frequencies (as in microsatellite loci in the human genome). This work considers the general case of mixed variables (continuous, binary and qualitative). The proposed method uses a variant of Rao's diversity coefficient [36], and develops an approach, to cluster units in the general situation of mixed data and replicates experiments. In [37] and [38] a first approximation for the algorithm in two particular situations, one continuous variable with and without replication, respectively can be found.

The remainder of this work is organized as follows. The previous concepts and notations are described in Section 2. The approach is formulated in Section 3. In Section 4, we present the results of a simulation study to illustrate some properties of the methodology. The results derived from the application of our procedure to synthetic and real data are presented in Section 5 and Section 6, respectively. Finally, we provide conclusions in Section 7.

2. Fundamentals

The starting point for most genetic or biomedical studies is a $n \times p$ data matrix $\mathbf{Z}$ where the rows correspond to the units to be clustered (genetic populations, genes...) and the columns to the variables to be measured (samples, chromosomal inversions, gene expression in different cellular time...).

First, we present a variant of Rao's diversity coefficient [36]. So, let C a group of n individuals which is represented by a p-random vector $\mathbf{Z} = (\mathbf{Z}_1, \ldots, \mathbf{Z}_p)$ with values on same metric space $\mathcal{R} \subset R^p$ and probability density f with respect to a suitable measure λ. Let $\mathbf{z}_i$, $i = 1, \ldots, n$, represent the individuals and suppose that a distance δ [39] is available between them. The geometric variability of C with respect to δ was defined as,

$$V_\delta(C) = \frac{1}{2} \int_{\mathcal{R}\times\mathcal{R}} \delta^2(\mathbf{z}_i, \mathbf{z}_j) f(\mathbf{z}_i) f(\mathbf{z}_j) \lambda(d\mathbf{z}_i) \lambda(d\mathbf{z}_j).$$

This quantity is a variant of Rao's diversity coefficient [36]. When δ is the Euclidean distance, $V_\delta(C) = tr(\text{VAR}(\mathbf{Z}))$. For other dissimilarities, $V_\delta(C)$ is a general measure of dispersion of $\mathbf{Z}$.

In applied problems the distance function is typically a datum, but the probability distribution for each population are unknown. A natural estimator given samples $\mathbf{z}_1, \ldots, \mathbf{z}_n$ is:

$$\hat{V}_\delta(C) = \frac{1}{2n^2} \sum_{l,m} \delta^2(\mathbf{z}_l, \mathbf{z}_m)$$

See [40] and references therein for a review of this concept, different properties and proofs. Recently, in [41] this concept has been applied in the context of the determination of the number of clusters and atypical units.

In [37] and [38] the so-called path distance was introduced. The definition of this distance was motivated by a hypothetical situation as in Figure 1. It seems very clear to consider two clusters $C_1 = \{1, 2, 3, 4, 5\}$ and $C_2 = \{6, 7\}$. But considering the Euclidean distance, the distance between objects 1 and 5 is the same that the distance between objects 5 and 7. So it seems natural to cluster together objects 1 and 5 as there are some objects between them and in some way joining them.

Figure 1. Hypothetical situation which motivated the definition of path distance.

This new distance needs the following definitions.

Definition 1. Given two elements i, j of set C, a path between i and j with respect to C, denoted P_{ij}^C, is a sequence $i = e_0 \rightarrow e_1 \rightarrow \ldots \rightarrow e_{t-1} \rightarrow e_t = j$, where $e_h \in C$, $h = 0, \ldots, t$. Two consecutive elements in the sequence defining a path P_{ij}^C, will be denoted as $(r, s) \in P_{ij}^C$.

Definition 2. Given two elements $i,\ j$ of set C and a fixed path between $i,\ j$ with respect to C, P_{ij}^C, the maximum step of the path is the maximum distance between two consecutive elements of the path, that is,

$$max\{d_{rs}\ :\ (r,s) \in P_{ij}^C\}.$$

Definition 3. Given two elements $i,\ j$ of set C, the path-distance between i and j with respect to C, $dp_C(i,j)$, is the minimum value of all maximum steps among all paths between i and j with respect to C, that is,

$$dp_C(i,j) = min_{P_{ij}^C}\{max(d_{rs}\ :\ (r,s) \in P_{ij}^C)\}.$$

Definition 4. Given a set C, the path-distance of C is the maximum of the path-distances between all pairs of elements $i,\ j$ of C,

$$dp(C) = max_{i,j \in C}\{dp_C(i,j)\}.$$

3. The Procedure

Consider now the general situation of mixed variables, where the experiment has been replicated R times.

Let C a group of n individuals which is represented by a p-random vector $\mathbf{Z} = (\mathbf{Z}_1, \ldots, \mathbf{Z}_p)$ with values on same metric space $\mathcal{R} \subset R^p$ and probability density f with respect to a suitable measure λ. Let z_{ir} be the value for individual i in replicate r, $i = 1, ..., n$, $r = 1, ..., R$. Then, for each replicate, there is one matrix $\mathbf{Z_r}$ containing the measurements of the n individuals for the p mixed variables.

Definition 5. Let i be an individual and $\mathbf{z}_{i,1}, \ldots, \mathbf{z}_{i,R}$ the corresponding measures for the R replicates. Let $\delta_{rs,\,i}$ be the distance between the r and s replicates of individual i. We define the *variability* of object i for the R replicates as:

$$DV_i = V_i^{1/2} = \left[\frac{1}{2R^2} \sum_{r,s=1}^{k} \delta_{rs,\,i}^2 \right]^{1/2}.$$

Before introducing the new algorithm, we need another definition.

Definition 6. Given a set C of individuals and a threshold value α, two clusters C_1 and C_2, $C_1 \cap C_2 = \emptyset$, $C_1 \cup C_2 = C$, define a partition of C feasible for λ and α if:

1. $dp(C_1) \leq \lambda$ and $dp(C_2) \leq \lambda$.
2. $\max\{|DV_i - DV_j|\ :\ i,j \in C_s\} \leq \alpha,\ s = 1,2.$

The first condition try to find compact clusters based in the path distance and the second one try to find compact clusters based on the variability between replicates. The partitions verifying both conditions, join similar individuals in relation with their distance and their variability between replicates. The second condition can be reformulating as follows:

Let $v_{min} = \min\{DV_i : i \in C\}$ and $v_{max} = \max\{DV_i : i \in C\}$ be the minimum and maximum variability of the objects of set C, respectively. Since (C_1, C_2) defines a partition of C, condition 2 is equivalent to:

2'. $DV_i \in [v_{min}, v_{min} + \alpha], \forall i \in C_1$ and $DV_i \in [v_{max} - \alpha, v_{max}], \forall i \in C_2$.

The aim of the following algorithm is to find the minimum value of λ such that there exists a partition of group C feasible with α and λ.

If the values of α and λ are fixed, it is easy to determine if there exist two clusters C_1 and C_2, $C_1 \cap C_2 = \emptyset$, $C_1 \cup C_2 = C$, determining a feasible solution with α and λ. Let $i_0 \in C$ be an individual selected at random in C verifying $DV_{i_0} \in [v_{min}, v_{min} + \alpha]$. Define cluster A that initially only contains this individual i_0. Then all the individuals i of C at distance from i_0 smaller than or equal to λ and verifying $DV_i \in [v_{min}, v_{min} + \alpha]$ are included in A. This process is repeated to include in A all the individuals of C with variability in $[v_{min}, v_{min} + \alpha]$ and at distance, for any other individual previously included in A, smaller than or equal to λ. The process terminates when no new individuals can be incorporated to A. Note that this process ensures that $dp(A) \leq \lambda$. If $A = C$, let j designate the last individual included in A. Then, clusters $C_1 = A \setminus \{j\}, C_2 = \{j\}$ define a feasible partition for α and λ of C. Otherwise, let $C_1 = A$ and define a new cluster B in a similar way as before, i.e., initialize cluster B with any arbitrary selected individual in $C \setminus C_1$ with variability in $[v_{max} - \alpha, v_{max}]$. Then, incorporate to B all individuals in $C \setminus C_1$ with variability in $[v_{max} - \alpha, v_{max}]$, at distance smaller than or equal to λ from any individual previously included in B and repeat the process until no more individuals can be added.

If $B = C \setminus C_1$, then, clusters C_1 and $C_2 = C \setminus C_1$ define a partition of C feasible for α and λ. Otherwise, it is not possible to determine a feasible partition.

Throughout, we will say that the response of the above procedure for the given values of λ and α is YES whenever it exists a feasible partition of C feasible for λ and α; and the response is NO when such partition does not exist.

Given a set C of individuals and a threshold value α, the optimal solution λ^* to the optimization problem of finding the smallest value of λ such that there exist a partition of C feasible for α and λ, is obtained by applying the so-called binary search technique (see for details [37]).

Then, we can present the proposed algorithm in the following way:

Initially: $t = 1$, $a_1 = \min_{i,j \in C} \delta_{ij}$, $b_1 = \max_{i,j \in C} \delta_{ij}$

While $a_t \neq b_t$ *do interactions*

In each one:

$\lambda_t := (a_t + b_t)/2$

Apply the first procedure with the value λ_t
If the response is YES *then*
$a_{t+1} = a_t$
$b_{t+1} = \max\{\delta_{ij} : \delta_{ij} \leq \lambda_t, i, j \in C\}$
Write C_1 y C_2
$\lambda^* = \lambda_t$
Otherwise
$a_{t+1} = \min\{\delta_{ij} : \delta_{ij} \geq \lambda_t, i, j \in C\}$
$b_{t+1} = b_t$
End while
Next:
$t := t + 1$
Finish when a_t=b_t

As commented in [37] for some particular cases, this optimization problem is no feasible or can be solved by other methods:

(a) When it exits an individual $i \in C$ such that $v_{min} + \alpha < DV_i < v_{max} - \alpha$, then the optimization problem is not feasible.

(b) When the problem is feasible and $v_{min} + \alpha \leq v_{max} - \alpha$, the optimization problem is trivial since $[v_{min},\ v_{min} + \alpha]$ and $[v_{max} - \alpha,\ v_{max}]$ define the only feasible partition for α and λ of C.

(c) When α is a value that $v_{min}+\alpha \geq v_{max}$, condition 2 is redundant and the feasibility of the partition depends only of the path distance λ. In this particular, the algorithm uses condition 1 as the unique condition and then it will be only based in the minimization of the corresponding path distances. In this case, the algorithm is equivalent to the Minimum Spanning Tree (MST) method (see [26] and [28]). However, when condition 2 is not redundant and the intervals $[v_{min},\ v_{min} + \alpha]$ and $[v_{max} - \alpha,\ v_{max}]$ overlap, the MST method is not able to solve the optimization problem, since there is no guarantee that squared root of the geometric variability between all the individuals in the resulting subtrees belong to the same interval.

(d) Note that the corresponding clustering is not necessarily unique. If $C = A$ the returned partition will depend on the initial parameters and in the case where $v_{max} - \alpha \leq v_{min} + \alpha$, the algorithm will return the biggest cluster for C_1.

For all these reasons, we recommend to fix the α value as $\alpha = A\,(v_{max} - v_{min})$ in such a way that $1/2 < A \leq 1$.

Finally, note that when the procedure is applied to no replicated experiments or when the value of α is $\alpha = v_{max} - v_{min}$, it generates successive divisions where each one has associated a value of λ in such a way that $dp(C_l) \leq \lambda$ $(l = 1, 2)$. So, the output can be represented as a binary tree where the high of the edges of a partition is given by the value of the path distance associate to each cluster of the partition. Moreover, the procedure leads to an associated ultrametric distance. However, when the experiment has been replicated and $\alpha = A\,(v_{max} - v_{min})$ with $1/2 < A \leq 1$, the obtained hierarchy will be not necessarily an indexed hierarchy.

4. Simulation Results

Several hierarchical clustering techniques have a tendency to cluster together, at relatively low level, individuals linked by a series of intermediates. This property, known as chaining may cause the methods to fail when there are a small number of individuals lying between relatively distinct clusters. On the other hand, it is well known that some methods tend to find spherical clusters and it is also useful to known the behavior of the clustering methods in front of scattered points. To study the behavior of the method in front of these questions, it was applied to different simulated datasets.

Spherical clusters. The procedure was applied to units produced by sampling twenty five or forty observations from each of two bivariate normal populations with mean vectors $\mu_1 = (0,0)$ and $\mu_2 = (4,4)$ and common covariance matrix Σ, where the variances are selected at random from $3^2, 4^2, 5^2$ and $0.25^2, 0.5^2, 0.75^2$, respectively. The correlation coefficient was selected at random from $0.2, 0.5, 0.8$. The data contain two well separated elliptical groups as it is shown in the Figure 2 (left). Only for 4 datasets the algorithm identify the two clusters correctly. For 8 datasets only one object was wrongly classified. That is, in the first step of the algorithm one object, object i was separated from the others obtaining two clusters i and $C \setminus \{i\}$. It was in the second step that the remaining objects were separated correctly as it is shown in Figure 2 (right). Finally in 5 datasets, 3 and 6 objects were incorrectly classified and for the other 4 datasets it was impossible to find the two initial groups.

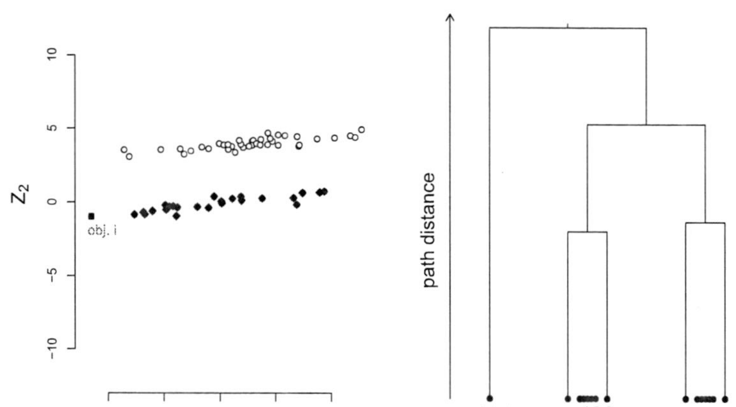

Figure 2. (Left) Graphic representation of one of the 20 sets of objects belonging to two groups C_1 (black points) and C_2 (white points) very well differentiated. (Right) Tree obtained by the cluster procedure. Note that object i in group C_1 (black points) was separated in the first step of the algorithm. In the second step, the other objects of C_1 and C_2 were well separated.

Chaining effect. The procedure was applied to units produced by sampling fifteen observations from each of two bivariate normal populations with mean vectors

$\mu_1 = (0,0)$ and $\mu_2 = (4,4)$ and common covariance matrix Σ, equal to the identity matrix. A small number of individuals lying between the groups were included. As it was expected, in any of the 20 generated datasets the algorithm correctly identified the initial groups. This was an expected result as the method is based in the path distance which considers two objects i and j similar if there are other objects between and near them.

Scattered points. Twenty datasets were generated (Figure 3). For each dataset and following [42], 10 two-dimensional groups were generated under the normal distribution with covariance matrices 0.1^2I, 0.2^2I,...,1^2I, respectively, where I is the identity matrix. The center of each group was selected at random in the $(0,\ 50) \times (0,\ 50)$ squared. All the groups have 25 objects in such a way that the distance between each object and the center of their group was minor than two times the standard deviation. From a uniform distribution in the $(0,\ 50) \times (0,\ 50)$ squared, 30 scattered new objects were generated. The distance between any of these scattered objects to the center of any of the 10 groups, was bigger than 3 times the corresponding standard deviation. The results showed that for the 20 datasets, the ten clusters and all the scattered points were well identify.

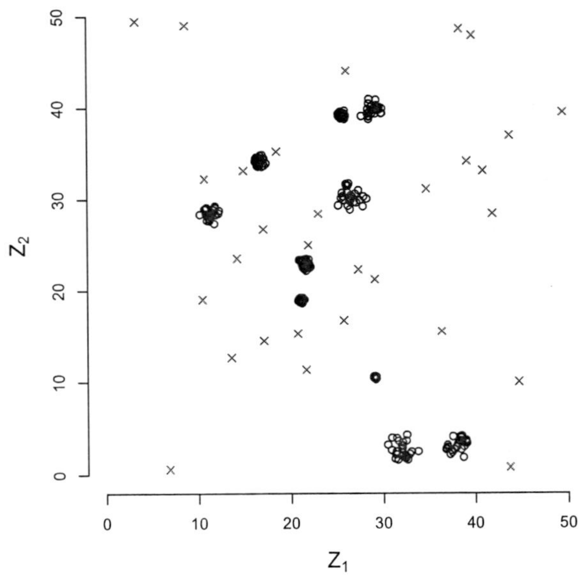

Figure 3. Position of ten well separated clusters with 30 scattered points (labeled by $\times$).

5. Synthetic Data

Continuous data. The following set of synthetic data [37] represents the expression of eleven genes measured in a microarray experiment. For each gene, three biological replicates are considered. These replicates may represent independent measurements on the same slide or measurements of the same genes in different slides. In this context, as it is usual, for a gene i was supposed that each replicate follows a $N(\mu_i, \sigma_i^2)$ distribution. The values of μ_i were obtained at random from a uniform distribution $U(0, 110)$. In order to generate genes with different variability between replicates, the σ_i values were obtained at random using the $U(0, 0.0001)$ and $U(5, 10)$ distributions for 5 and 6 genes, respectively. The generated values are presented in Table 1.

Table 1. Synthetic data set representing eleven genes and three replicates. The last two columns show the mean and standard deviation values.

Genes	Replicate 1	Replicate 2	Replicate 3	Mean	Standard deviation
1	3,88	3,78	5,53	4,40	0,98
2	3,01	7,34	1,40	3,92	3,07
3	17,36	16,64	18,86	17,62	1,13
4	14,10	20,28	10,31	14,90	5,03
5	8,32	10,35	1,56	6,74	4,60
6	33,65	37,11	44,80	38,52	5,71
7	37,66	42,06	43,47	41,06	3,03
8	44,98	44,83	45,64	45,15	0,43
9	44,97	42,13	42,86	43,32	1,47
10	45,56	45,45	50,31	47,11	2,78
11	60,72	60,25	59,93	60,30	0,40

The Euclidean distance was used to measure the distances between objects. Figure 4 shows the obtained results for $\alpha = 2/3(v_{max} - v_{min})$.

For simplifying, Figure 4 does not show the partitions where each cluster contains only two objects. The obtained clusters are:

$$C_1 = \{1, 2, 3, 7, 8, 9, 10, 11\}, C_2 = \{4, 5, 6\}, C_3 = \{1, 3, 8, 9, 11\},$$
$$C_4 = \{2, 7, 10\}, C_5 = \{4, 5\}, C_6 = \{6\}, C_7 = \{1, 3, 9\}, C_8 = \{8, 11\},$$
$$C_9 = \{2\}, C_{10} = \{7, 10\}, C_{11} = \{9\}, C_{12} = \{1, 3\}.$$

Note that the algorithm takes into account the mean expression level for the clusters and the variability between their replicates. For example, the first division separates the genes with large variability between replicates (cluster $C_2 = \{4, 5, 6\}$). In this step the clusters are very homogenous with respect to the expression level, as indicates the value of the path distances $dp(C_2) = 43.07$ and $dp(C_1) = 40.79$. On the other hand, when we considered cluster $C_4 = \{2, 7, 10\}$ and its division in clusters $C_9 = \{2\}$ and $C_{10} = \{7, 10\}$, we can observed that even genes 2 and 7 have similar variability between replicates, these two genes were separated because their expression levels are very different. To make the classification of genes considering these two questions will be very useful, as it permits to determine which genes have similar variability between replicates and also at which mean expression level their are functional.

General case. Now, we consider the general case of an experiment with replicates and no necessarily continuous variables. Two very well differentiated groups, one of them with high variability between replicates, were generated. Each group had 20 objects and was generated from the multinomial distribution in the following way. Two groups in $m = 3$

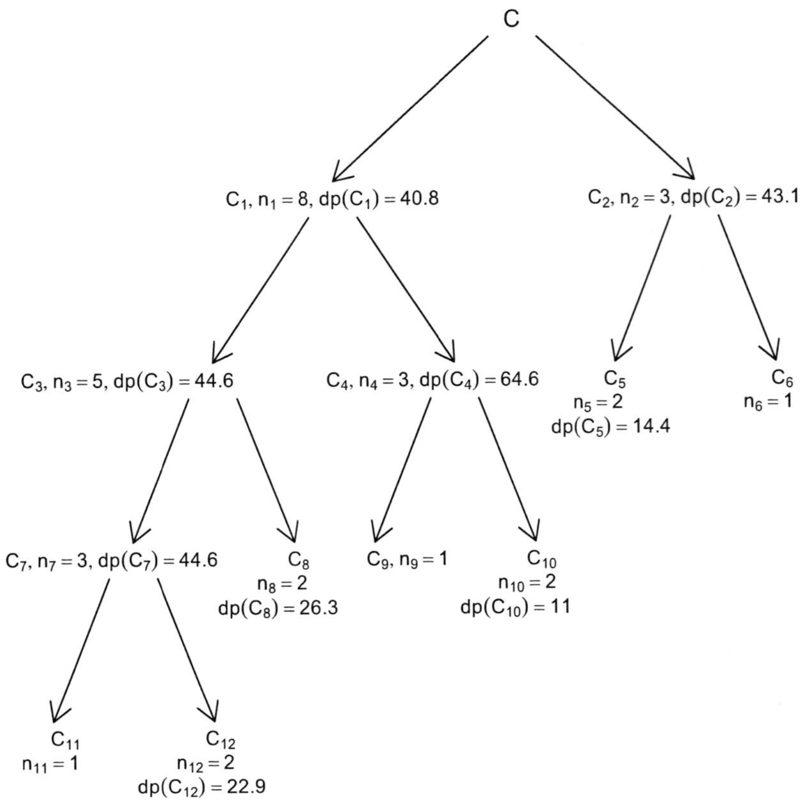

Figure 4. Graphic representation of the obtained clusters using a synthetic data set representing eleven genes and three replicates. In each step the number of objects in each cluster and the associated path distanced are showed.

dimensions: group $C_1 \sim M(100, \pi_1 = 0.8, \pi_2 = 0.1, \pi_3 = 0.1)$ (objects $1-20$) and group $C_2 \sim M(10, \pi_1 = 0.1, \pi_2 = 0.5, \pi_3 = 0.4)$ (objects $21 - 40$). For each object, $R = 3$ replicates were generated. Note that the variability between replicates in group C_2 is bigger than in C_1. Each object i is characterized by an m-dimensional vector $(p_{i1}, ..., p_{im})$ in the *space* whose coordinates are the generated frequencies with

$$p_{il} \geq 0,\ l = 1, ..., m \text{ and } \sum_{l=1}^{m} p_{il} = 1; \quad i = 1, ..., n.$$

Mathematically, this space constitutes a manifold with boundary of dimension $m - 1$. We can define a distance between two units $i = (p_{i1}, ..., p_{im})$ and $j = (p_{j1}, ..., p_{jm})$ by

$$d(i, j) = \arccos \sum_{l=1}^{m} \sqrt{p_{il} p_{jl}},$$

where $d(i, j)$ is the Bhattacharyya distance [43].

The solutions obtained for $\alpha = 4/5(v_{max} - v_{min})$ and $\alpha = 2/3(v_{max} - v_{min})$, respectively are shown in Figure 5. For $\alpha = 4/5(v_{max} - v_{min})$, the correct partition is obtained at the first division. For $\alpha = 2/3(v_{max} - v_{min})$, the first partition ($\widetilde{C_1} = \{1, \ldots, 20, 21, 22, 23, 24, 25, 27, 28, 30, 31, 37, 38, 39\}$ and $\widetilde{C_2} = \{26, 29, 32, 33, 34, 35, 36, 40\}$) separates one cluster ($\widetilde{C_2}$) containing the objects with higher variability between replicates from the other objects which are clustered in $\widetilde{C_1}$. This occurs as there are objects i in the second cluster such that $DV_i \in [v_{min}, v_{min} + \alpha]$, these objects are assigned at cluster $\widetilde{C_1}$ for the condition 2 and they act as intermediate objects when the rest of the objects from the second group are clustered in cluster $\widetilde{C_1}$.

It is clear that the α value is chosen by the investigator in such a way that if the most important is to identify genes with high variability between replicates, a small value of α will be considered and in other case, a big value of α will be selected.

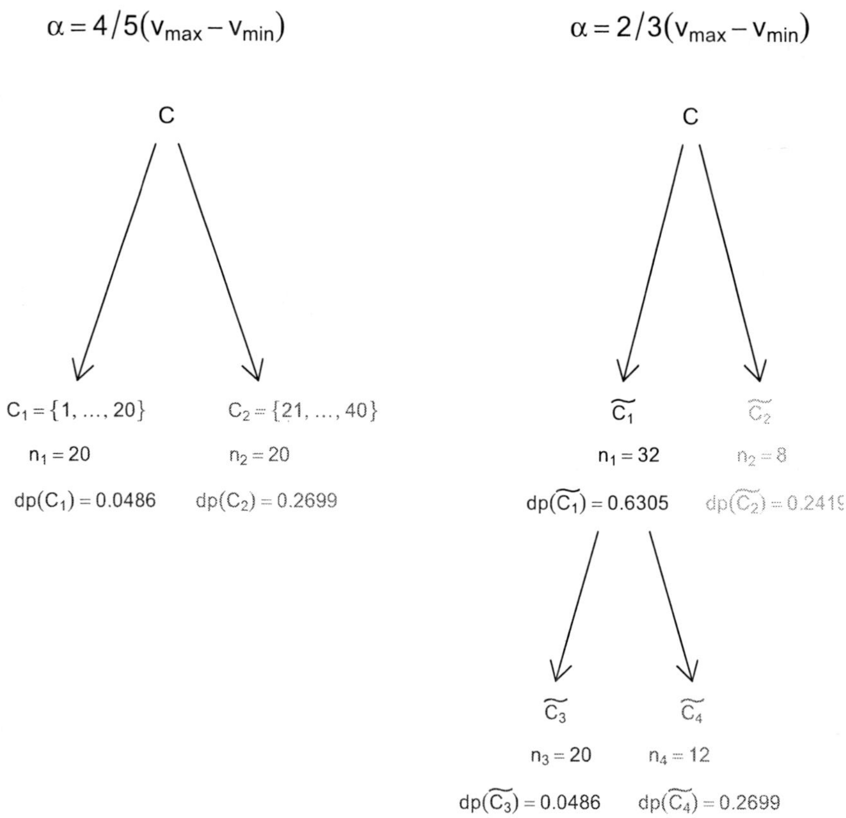

Figure 5. (left) Graphic representation of the obtained clusters using the value $\alpha = 4/5(v_{max} - v_{min})$ on a synthetic data set with three replicates and no continuous data. (right) Graphic representation of the obtained clusters using the value $\alpha = 2/3(v_{max} - v_{min})$ on a synthetic data set with three replicates and no continuous data. In both cases and for each step, the number of objects in each cluster and the associated path distance are showed.

6. The Procedure on Real Data

The procedure was applied to five previously published datasets. One for biomedical data related to diagnosis of dermal diseases [44] and four gene-expression data, the leukemia (ALL/AML) dataset of [45], the cutaneous melanoma dataset of [46], the cholesterol data set of [47] and the aging in *C. elegans* dataset of [48]. The first dataset corresponds to mixed variables without replication and the gene-expression datasets correspond to experiments with or without replicates and one observed continuous variable. As these last four datasets have been described in detail in [37] and [38] we choose only one of them for illustrating here this situation.

6.1. Dermal Diseases

We considered a subset of a public data set, in the medical domains of dermatology located in the data repository UCI Knowledge Discovery in Databases [44]. This data set was provided by H. A. Guvenir (Department of Computer Engineering and Information Science, Bilkent University, Turkey). The domain contains instances of patients with known diagnosis or classes. Here we consider three different diagnoses, and all of them share some clinical characteristics (with few differences) of erythema and scaling. The diseases are lichen planus, chronic dermatitis and pityriasis rubra pilaris. Some characteristics of the diseases are discussed in [49]. The considered data set contains 144 instances (72 lichen planus, 52 chronic dermatitis and 20 pityriasis rubra pilaris) presenting 34 different clinical attributes. For each instance, 12 clinical features (i.e. age, family history and so on) and 22 histopathological features (i.e. melanin incontinence, eosinophils in the infiltrate, exocytosis, hyperkeratosis and so on) obtained after a biopsy of a patient skin sample are available. We used the Gower's distance [50]. This distance (squared) is obtained by $d_{ij}^2 = 2(1 - s_{ij})$, where s_{ij} is defined as:

$$s_{ij} = \frac{\sum_{l=1}^{m_1}\left(1 - \frac{|x_{il} - x_{jl}|}{R_l}\right) + a + \alpha}{m_1 + (m_2 - d) + m_3}$$

where R_l is the range of the lth continuous variable ($l = 1, ..., m_1$); for the m_2 binary variables, a and d represent the number of matches presence-presence and absence-absence, respectively and α is the number of matches between states for the m_3 qualitative variables. The algorithm (see Figure 6) separated in the first step, one cluster C_1 containing only one lichen planus instance. In the second step, cluster C_2 was separated in one cluster (cluster C_3) with all the remaining lichen planus instances and cluster C_4 with the other instances. In the next step, C_4 was separated in cluster C_5 with one pityriasis rubra pilaris instance and cluster C_6 with the remaining instances. Similar results were obtained in the next two divisions, with only one pityriasis rubra pilaris instance being separated of the rest. Finally, C_{10} was separated in two new clusters, one with all the remaining pityriasis rubra pilaris instances, and the other with all the chronic dermatitis instances. Note that the procedure provides a good classification.

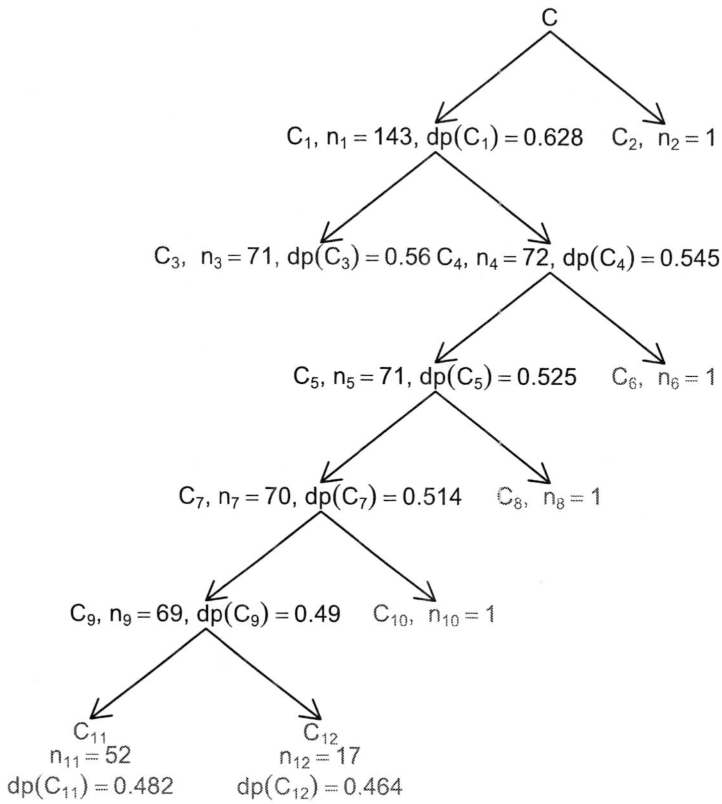

Figure 6. Graphic representation of the obtained clusters with the dermal disease data. In each step, the number of objects in each cluster and the associated path distanced are showed.

6.2. Cholesterol Data

During more than fifty years, the great inverse relation between cholesterol HDL (High Density Hipoprotein) and the arteriosclerosis has been recognize. However, the mechanisms related with this function are not well establish. The use of transgenic and knockout mice have been very useful. For example, the knockout mice for the apolipoprotein AI (ApoAI) present very low HDL cholesterol levels [51], [52]. Here, we use the data of [47], where using a replicated microarray experiment, they studied the expression level of the gene which controls the ApoAI in the liver of knockout and control mice. The authors wanted to understand the effects of ApoAI on the expression of other genes at the liver level. We used a subset of the data in [47]. The analysis is based on expression levels for the top 50 genes that are differentially expressed according to the ranked t statistic presented in [47] across three replicates. This row-column design with 50 rows and three columns is presented in [37]. One gene (gene 2) presented a high standard deviation in the expression level between replicates (0.95) respect to the others (maximum value = 0.58), and for this reason it was excluded.

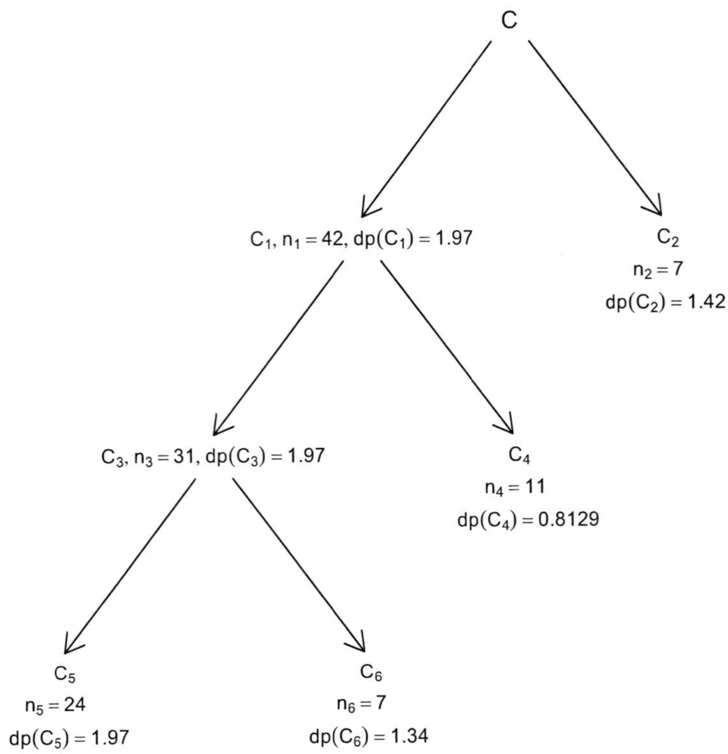

Figure 7. Graphic representation for the obtained clusters on the microarray experiment on functional mechanisms of the HDL cholesterol data. For each step the number of objects in each cluster and the associated path distanced is showed.

The algorithm, taking $\alpha = 2/3(v_{max} - v_{min})$, gave in the first step two clusters (see Figure 7), one of them including 7 genes with great variability between replicates (cluster C_2) and the other cluster (cluster C_1) containing the 42 remaining genes. The former cluster has genes with a high variability in gene expression, thus few important for the studied phenotype. In the next step of the algorithm (partition of cluster C_1), the 11 genes with lower variances were separated of the rest (cluster C_4). These genes have a constant level of expression. Some presented high mean (over expression) while others have low means (under expression). In the expression of the trait, the former group is constituted by active genes while the later by repressed genes. The remaining cluster (cluster C_3) with 31 genes was analyzed in the same way. Then, a cluster with 24 genes with low variance between replicates was obtained (cluster C_5), thus concluding that they are rather constant in gene expression and thus important for the phenotype. The other cluster (cluster C_6) with 7 genes, is composed by genes with rather high variability, thus no very important for the analyzed trait.

As shown in this example, the algorithm gives higher priority to the between replicate variability. At each partition, the algorithm tends to return a cluster of genes with low variance across replicates and other with high variance across replicates, indicating genes with variable level of expressions. Only later does the criterion relating to the path distance start to have much impact, so the criterion relating to the between-replicates variability is given higher priority.

7. Conclusion

Our methodology is general and enjoys several desirable properties. First, it does not require continuous variables or any other restriction on the data. Second, it gives a rigorous and useful treatment of the replicate experiments, considering both the mean expression profile and the variability of expression.

In conclusion, we believe that methods of clustering that take into account the replicates experiments and are applicable to any kind of data will become increasingly valuable for genetic and biomedical situations.

References

[1] Eisen, MB; Spellman, P; Brown, PO; Botstein, D. Cluster analysis and display of genome-wide expression patterns, *Proceedings of the National Academy of Sciences of the United States of America*, 1998, vol. 95, pp. 14863-14868.

[2] Wen, X; Fuhrman, S; Michaels, GS; Carr, DB; Smith, S; Barker, JL; Somogyi, R. Large-scale temporal gene expression mapping of central nervous system development, *Proceedings of the National Academy of Sciences of the United States of America*, 1998, vol. 95, pp. 334–339.

[3] Hartigan, JA; Wong, MA. A k-means clustering algorithm, *Applied Statistics*, 1979, vol. 28, pp. 126–130.

[4] Herwig, R; Poustka, AJ; Meuller, C; Lehrach, H; O'Brien, J. Large-scale clustering of cDNA–fingerprinting data, *Genome Research*, 1999, vol. 9, pp. 1093–1105.

[5] Tavazoide, S; Hughes, J; Campbell, M; Cho, RJ; Churo, GM. Systematic determination of genetic network architecture, *Nature Genetics*, 1999, vol. 22, pp. 281–285.

[6] McLachlan, GJ; Basford, K. *Mixture models: inference and applications to clustering*. Marcel Dekker. New York; 1988.

[7] Fraley, C; Raftery, A. Model-based clustering, discriminat analysis, and density estimation, *The Computer Journal*, 2002, vol. 97, pp. 611–631.

[8] Kohonen, T. The self-organizing map, *Proceedings IEEE*, 1990, vol. 78, pp. 1464–1479.

[9] Kaufman, L; Rousseeuw, PJ. *Finding groups in a data*. Wiley. New York; 1990.

[10] Tamayo, P; Slonim, D; Mesirov, J; Zhu, Q; Kitareewan, S; Dmitrovsky, E; Lander, ES; Golub, TR. Interpreting patterns of gene expression with self-organizing maps: Methods and application to hematopoietic differentiation, *Proceedings of the National Academy of Sciences of the United States of America*, 1999, vol. 96, pp. 2907–2912.

[11] Sharan, R; Shamir, R. CLICK: a clustering algorithm with applications to gene expression analysis, *Proceedings ISMB*, 2000, pp. 307–316.

[12] Lukashin, A; Fuchs, R. Analysis of temporal gene expression profiles: clustering by simulated annealing and determining the optimal number of clusters, *Bioinformatics*, 2001, vol. 17, pp. 405–414.

[13] Yeung, K; Haynor, D; Ruzzo, W. Validating clustering for gene expression data, *Bioinformatics*, 2001, vol. 17, pp. 309–318.

[14] Fraley, C; Raftery, AE. MCLUST: Software for Model-Based Clustering, Discriminant Analysis and Density Estimation, Technical Report no. 415 (2002), Department of Statistics, University of Washington.

[15] McLachlan, GJ; Bean, RW; Peel, D. A mixture model-based approach to the clustering of microarray expression data, *Bioinformatics*, 2002, vol. 18, pp. 1–10.

[16] Steuer, R; Kurths, J; Daub, C; Weise, J; Selbig, J. The mutual information: detecting and evaluating dependencies between variables, *Bioinformatics*, 2002, vol. 18, pp. 231–240.

[17] Hastie, T; Tibshirani, R; Friedman, J. *The elements of statistical learning: data mining, inference and prediction*. Springer. New York; 2002.

[18] Sharan, R; Maron-Katz, A; Shamir, R. Clik and expander: a system for clustering and visualizing gene expression data, *Bioinformatics*, 2003, vol. 19, pp. 1787–1799.

[19] Dembele, D; Kastner, P. Fuzzy c-means method for clustering microarray data, *Bioinformatics*, 2003, vol. 19, 973–980.

[20] Horn, D; Axel, I. Novel clustering algorithm for microarray expression data in a truncated svd space, *Bioinformatics*, 2003, vol. 19, pp. 1110–1115.

[21] Qin, J; Lewis, D; Noble, W. Kernel hierarchical gene clustering from microarray gene expression data, *Bioinformatics*, 2003, vol. 19, pp. 2097–2104.

[22] Dhilon, I; Marcotte, E; Roshan, U. Diametrical clustering for identifying anticorrelated gene clusters, *Bioinformatics*, 2003, vol. 19, pp. 1612–1619.

[23] Dudoit, S; Fridlyand, J. Bagging to improve the accuracy of a clustering procedure, *Biometrics*, 2003, vol. 19, pp. 1090-1099.

[24] Luo, F; Khan, L; Bastani, F; Yen, IL; Zhou, J. *Finding Groups in Data: an Introduction to Cluster Analysis*. Wiley. New York; 2004.

[25] Dugas, M; Merk, S; Breit, S; Dirschedl, P. Mdclust: Exploratory microarray analysis by multidimensional clustering, *Bioinformatics*, 2004, vol. 20, pp. 931–936.

[26] Xu, Y; Olman, V; Xu, D. Clustering gene expression data using a graph-theoretic approach: an application of minimum spanning trees, *Bioinformatics*, 2002, vol. 18, pp. 536–545.

[27] Xu, Y; Olman, V; Wang, L; Xu, D. Clustering EXCAVATOR: a computer program for efficiently mining gene expression data, *Nucleic Acids Research*, 2003, vol. 31, pp. 5582–5589.

[28] Varma, S; Simon, R. Iterative class discovery and feature selection using Minimal Spanning Trees, *BMC Bioinformatics*, 2004, vol. 5, pp. 126–134.

[29] Sherlock, G. Analysis of large-scale gene expression data, *Current Opinion Immunology*, 2000, vol. 12, pp. 201–205.

[30] Lee, MT; Kuo, FC; Whitmore, GA; Sklar, J. Importance of replication in microarray gene expression studies: Statistical methods and evidence from repetitive cDNA hybridizations, *Proceedings of the National Academy of Sciences of the United States of America*, 2000, vol. 97, pp. 9834–9839.

[31] Hughes, TR; Marton, MJ; Jones, AR; Roberts, CJ; Stoughton, R; Armour, CD; Bennett, HA; Coffey, E; He, YD. Functional discovery via a compendium of expression profiles, *Cell*, 2000, vol. 102, pp. 109–126.

[32] Yeung, KY; Medvedovic, M; Bumgarner, RE. Clustering gene expression data with repeated measurements, *Genome Biology*, 2003, vol. 4, R34.

[33] Medvedovic, M; Yeung, KY; Bumgarner, RE. Bayesian mixture model based clustering of replicated microarray data, *Bioinformatics*, 2004, vol. 8, pp. 1222–1232.

[34] Ridder, D; Staal, F; van Dogen, JM; Reinders, MJ. Maximum significance clustering of oligonucleotide microarrays, *Bioinformatics*, 2006, vol. 22, pp. 326–331.

[35] Sudbery P. *Human Molecular Genetics*. Prentice-Hall. Harlow. United Kindom; 2002.

[36] Rao, CR. Diversity: its measurement, decomposition, apportionment and analysis. *Sankhyā. The Indian Journal of Statistics, Series A*, 1982, vol. 44, pp. 1-22.

[37] Irigoien, I; Fernández, E; Vives, S; Arenas, C. Clapper: A clustering algorithm based on path-distances for experimental replicates, *Advances and Applications in Statistics*, 2006, vol. 6, pp. 379–400.

[38] Irigoien, I; Fernández, E; Vives, S; Arenas, C. Clum: A cluster program for analyzing microarray data, *Russian Journal of Genetics*, 2008, vol. 44, pp. 993–996.

[39] Gower, JC. *Measures of similarity, dissimilarity and distance.* In: *Encyclopedia of Statistical Sciences*, volume 5. New York: John Wiley and Sons; 1985; pp. 397–405.

[40] Arenas, C; Cuadras, CM. Some recent statistical methods based on distances, *Contributions to Science*, 2002, vol. 2, pp. 183–191.

[41] Irigoien, I; Arenas, C. INCA: New statistic for estimating the number of clusters and identifying atypical units, *Statistics in Medicine*, 2008, vol. 27, pp. 2948–2973.

[42] Tseng, GC. Tight clustering: A resampling-based approach for identifying stable and tight patterns in data, *Biometrics*, 2005, vol. 61, pp. 10–16.

[43] Bhattacharyya, A. On a measure of divergence of two multinominal populations, *Sankhyā*, 1946, vol. 7, pp. 401–406.

[44] Hettich, S; Bay, SD. The UCI KDD Archive (http://kdd.ics.uci.edu), 1999, Department of Information and Computer Science. University of California at Irvine, Irvine, CA.

[45] Golub, TR; Slonim, DK; Tamayo, P; Huard, C; Gaasenbeek, M. et al., Molecular classification of cancer: Class discovery and class prediction by gene expression monitoring, *Science*, 1999, vol. 286, pp. 531–537.

[46] Bittner, M; Meltzer, P; Chen, Y; Jiang, Y; Seftor, E. et al., Molecular classification of cutaneous malignant melanoma by gene expression profiling, *Nature*, 2000, vol. 406, pp. 536–540.

[47] Callow, MJ; Dudoit, S; Gong, EL; Speed TP; Rubin, EM. Microarray expression profiling identifies genes with altered expression in HDL deficient mice, *Genome Research*, 2000, vol. 10, pp. 2022–2029.

[48] Lund, J; Tedesco, P; Duke, K; Wang, J; Kim, SK; Hohnson, TE. Transcriptional profile of aging in *C. elegans*, 2002, *Current Biology*, vol. 12, pp. 1566–1573.

[49] Guvenir, HA; Demiroz, G; Ilter, N. Learning differential diagnosis of erythematosquamous diseases using voting feature intervals, 1998, *Artificial Intelligence in Medicine*, vol. 13, pp. 147-165.

[50] Gower, JC. A general coefficient of similarity and some of its properties, *Biometrics*, 1971, vol. 27, pp. 857–871.

[51] Williamson, R; Lee, D; Hagaman, J; Maeda, N. Marked reduction of high density lipoprotein cholesterol in mice genetically modified to lack apolipoprotein A-I, 1992, *Proceedings of the National Academy of Sciences of the United States of America*, vol. 89, pp. 7134–7138.

[52] Plump, AS; Erickson, SK; Weng, W; Partin, JS; Breslow, JL; Williams, DL. Apolipoprotein A-I is required for cholesteryl ester accumulation in steroidogenic cells and for normal adrenal steroid production, 1996, *The Journal of Clinical Investigation*, vol. 97, pp. 2660–2671.

Reviewed by F. Mestres (Associate Professor). Department of Genetics. University of Barcelona, Barcelona (Spain).

In: Cluster Computing and Multi-Hop Network Research ISBN 978-1-60876-186-9
Editors: C. Jimenez and M. Ortego, pp. 109-136

Chapter 6

Design of Scalable Parallel Algorithms for Large Scale Data Analysis and Visualization

***Qishi Wu*[1,*] *and Jinzhu Gao*[2,†]**
[1]Department of Computer Science
The University of Memphis, Memphis, TN 38152, USA
[2]Department of Computer Science
University of The Pacific, Stockton, CA 95211

Abstract

The rapid advances in computing power made possible by the modern supercomputing technologies have enabled a wide spectrum of next-generation scientific applications with a primary goal of either simulating very complex physical phenomena or creating digital representations of real-world objects for educational or scientific purposes. These applications generate vast amounts of scientific data on the order of terabytes at present or petabytes in the predictable future, which must be transferred, visualized, and analyzed by a collaborative team of geographically distributed scientists. The capability of collaboratively analyzing and visualizing today's abundant scientific data and on-going computations over wide-area networks is critical to ensure both the success of mission-critical applications and the utilization of expensive computing resources. In this chapter, we discuss various issues in the design of scalable parallel algorithms for large-scale data analysis and visualization and present a number of novel techniques with focus on scalability improvement. The main goal of algorithm design for parallel computing is to balance workload as well as minimize runtime communication and synchronization costs. We present several data management and distribution schemes that promise highly scalable parallel performance by utilizing both out-of-core schemes and effective data traversal along a space filling curve. These schemes ensure balanced workload for different parallel implementations and avoid data replication and run-time redistribution to maximize the utilization of limited system resources. We also present a linear pipeline scheme for image compositing in parallel visualization to support efficient image delivery to a remote client.

*E-mail address: qishiwu@memphis.edu
†E-mail address: jgao@pacific.edu

The linear pipeline scheme arranges an arbitrary number of parallel processors within a cluster in a linear order and divides the image into a carefully selected number of segments, which flow through the linear in-cluster pipeline and wide-area networks to the remote client consecutively. We analytically determine the segment size that minimizes the final image display time and derive the conditions where the proposed image compositing and delivery scheme outperforms the traditional scheme based on the binary swap algorithm. The efficacy of these approaches is demonstrated by case studies that involve data analysis and visualization techniques widely used in scientific fields.

1. Introduction

The rapid advances in computing power made possible by the modern supercomputing technologies have enabled a wide spectrum of next-generation scientific applications with a primary goal of either simulating very complex physical phenomena or creating digital representations of real-world objects for educational or scientific purposes. These applications generate vast amounts of scientific data on the order of terabytes at present or petabytes in the predictable future, which must be transferred, visualized, and analyzed by a collaborative team of geographically distributed scientists. Examples of such large-scale scientific applications include the Terascale Supernova Initiative (TSI) [1] project at Oak Ridge National Laboratory (ORNL), where astrophysicists from leading research institutes and universities in US develop core collapse models and run supernovae simulations on the National Leadership Computing Facilities (NLCF) [2]. With a minimal portion of the potential physics variables, the TSI simulation code currently generates one terabytes data on a daily basis .

The enormous data sizes pose unprecedented challenges to the data generation, analysis and visualization process that utilizes various system resources such as computing power, storage space, and network bandwidth, and have rendered the traditional computing platform based on a standalone computer inadequate to meet growing performance demands. In 2004, a community of scientific researchers identified limitations in data management and understanding as a primary bottleneck to scientific progress [25]. Computing-intensive tasks of ever-increasing complexity in these large-scale applications often require pooling globally-located resources to perform data collections, simulations, visualizations, and analysis in a parallel and distributed manner. It has been widely recognized that the utilization of high-performance computing resources offers a promising solution to support and enhance the analysis and visualization process of large scale data that is extremely difficult to carry out on a single personal computer.

The capability of collaboratively analyzing and visualizing today's abundant scientific data and on-going computations over wide-area networks is critical to ensure both the success of mission-critical applications and the utilization of expensive computing resources [23]. The on-line simulation monitoring and interactive computational steering through visual feedback constitutes an important part of these research processes [35]. To some extent, remote analysis and visualization has become an indispensable research tool for many large-scale scientific applications [15, 3]. The final images generated by visualization systems often need to travel through wide-area wired or wireless network connections to be viewed on the screen of a remote laptop, desktop, or powerwall [18, 32]. Such a remote analysis and visualization system generally consists of stages including data partition and distribution, data rendering, image compositing, and image delivery. Depending on the nature of objects to be visualized, data volume, number of processors, image size, network bandwidth as well as the algorithm designed for each individual task, the bottleneck of the entire process could occur at either one of these stages.

In this chapter, we discuss various issues in the design of scalable parallel algorithms for large-scale data analysis and visualization and present a number of novel techniques with focus on scalability improvement. The main goal of algorithm design for parallel computing is to balance workload as well as minimize runtime communication and synchronization costs. We present several data management and distribution schemes that promise highly scalable parallel performance by utilizing both out-of-core schemes and effective data traversal along a space filling curve. These schemes ensure balanced workload for different parallel implementations and avoid data replication and run-time redistribution to maximize the utilization of limited system resources. We also present a linear pipeline scheme for image compositing in parallel visualization to support efficient image delivery to a remote client. The linear pipeline scheme arranges an arbitrary number of parallel processors within a cluster in a linear order and divides the image into a carefully selected number of segments, which flow through the linear in-cluster pipeline and wide-area networks to the remote client consecutively. We analytically determine the segment size that minimizes the final image display time and derive the conditions where the proposed image compositing and delivery scheme outperforms the traditional scheme based on the binary swap algorithm. The efficacy of these approaches is demonstrated by case studies that involve data analysis and visualization techniques widely used in scientific fields.

2. Related Work

The size of a typical large scale dataset has continued to grow from gigabytes in the past few years to terabytes at present to petabytes in the near future. Datasets of such volumes pose a great challenge to traditional scientific visualization algorithms in various aspects of performance requirements on storage space, computation speed and network bandwidth. A great deal of research efforts have been focused on finding effective solutions to overcome these performance restrictions in large data visualization.

Among these solutions, parallel algorithms utilizing the computing power of parallel machines to improve the run-time interactivity of large data visualization stand out perhaps as the most promising options. Scalability is one of the critical performance metrics in evaluating these solutions. Molnar *et al.* [24] classified parallel polygon rendering algorithms into three categories: sort-first, sort-middle and sort-last, and presented a thorough comparison between and discussion on these three algorithms. A sort-first algorithm distributes raw primitives during geometry processing. A sort-middle algorithm redistributes all screen-space primitives after geometry processing and before rasterization. A sort-last algorithm redistributes pixels, samples or pixel fragments after geometry processing and part of rasterization. To take advantage of both sort-first and sort-last algorithms, Samanta *et al.* proposed a hybrid sort-first and sort-last parallel rendering algorithm on PC clusters [30].

Research efforts have also been devoted to the study of parallel algorithms for isosurface extraction. Hansen and Hinker [13] proposed a parallel algorithm on SIMD machines. Ellsiepen introduced a parallel implementation of unstructured isosurface extraction with a dynamical block distribution scheme [7]. Crossno and Angel proposed an isosurface extraction algorithm using particle systems and its parallel implementation [6] . An isosurface extraction algorithm in span space and its parallelization are proposed in [31].

Improving the performance of volume rendering is another focus area that attracts much attention from researchers. Ma *et al.* [22] proposed a parallel volume-rendering algorithm that distributes data evenly to available computing resources and produces final images using binary-swap compositing. A scalable volume rendering technique was presented in [19], which utilizes lossy compression and achieves interactive rendering of time-varying scalar datasets.

In recent years, the study of time-varying datasets has gained increasing interest in various science domains. To utilize parallel computer power in time-varying volume visualization, Rowlan [27] and Ma [20, 21] proposed to perform the tracking of time-varying data using direct volume rendering on a massively parallel computer. Some software systems, such as VISUAL3 [12] and SCIRun [26] also support the tracking of 3D datasets in a distributed computing environment.

3. Data Partition and Distribution

One of the main goals of a parallel algorithm is to achieve high scalability. Towards this goal, the algorithm needs to balance workload and minimize run-time communication overhead and synchronization costs. For large volume visualization, besides the efficiency of the underlying algorithm, the performance of a scalable parallel implementation also depends on the effectiveness in data management and distribution. In this chapter, for the sake of expandability, we use PC clusters as the targeted computing platform, where efficient and effective data management and distribution remain to be a tremendous challenge.

3.1. Basic Ideas

We will start with an introduction to four traditional methods for data partition and distribution as follows:

1. The data is replicated and stored at each node as shown in Fig. 1. Since each node accesses data on its local disk, the run-time data communication is completely eliminated. However, when dealing with a large scale data, this method may not be practically feasible. A k-way replication technique [29] was proposed to alleviate the storage limitation problem by replicating each 3D primitive of a large scene on k out of n PCs ($k \ll n$). This method still incurs the overhead of dynamic data redistribution at run time because some processors may not have a copy of the data in need. In addition, even duplicating a small portion of terabytes data requires a significant amount of storage space.

2. A single copy of the data is stored only on the host node, which is responsible for sending the required data to other nodes at run time, as shown in Fig. 2. This scheme

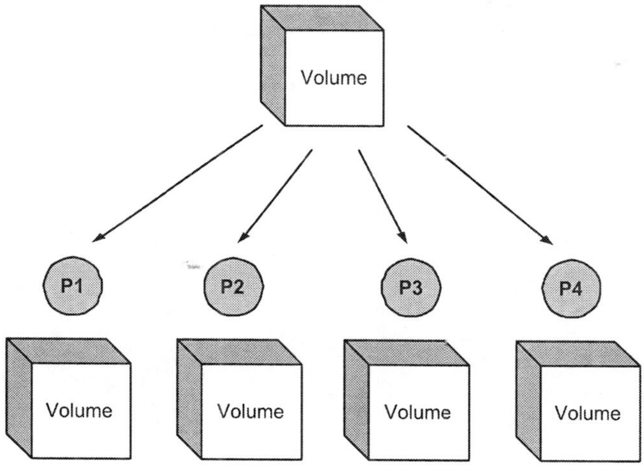

Figure 1. The volume is duplicated at each node.

avoids replicating the entire data at each node. However, the expensive cost for run-time data redistribution makes it unsuitable for large-scale data processing.

3. The data is partitioned into p chunks, which are distributed to p nodes on a one-to-one basis, as shown in Fig. 3. During the course of rendering, each node is only responsible for rendering the chunk that was assigned to it. This approach avoids data replication. However, the rendering workload of each chunk can be different due to the unpredictable distribution of interested area in the data, thus maintaining balanced workload would be challenging in this approach without incurring additional data communication cost.

4. The data is partitioned into n data blocks ($n \gg p$, p is the number of processing nodes) and each block is assigned to one node according to some distribution rules, as shown in Fig. 4. During the rendering process, each node is only responsible for rendering the blocks assigned to it. This method also avoids data replication. However, the quality of load balancing highly depends on the data distribution method we choose. Zhang *et al.* proposed a random data distribution scheme to achieve balanced workload and proved that a static data distribution method can also promise balanced workload [37].

Based on the analysis of the aforementioned traditional approaches, we proposed a data partition and distribution scheme that meets the following criteria to handle large scale data [9]:

- The scheme should avoid data replication to maximize the utilization of limited system resources.
- The scheme should be static to avoid expensive run-time data redistribution.
- The scheme should be general enough to ensure balanced workload for different parallel algorithms such as parallel view-dependent visualization and parallel multi-resolution volume rendering.

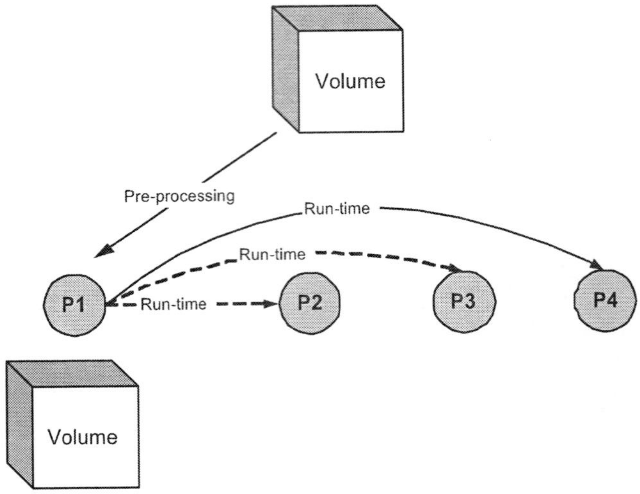

Figure 2. Only the host node has a copy of the volume and it is responsible for sending the required data to other nodes at run time.

In [9], we partition the entire data volume into a number of data blocks to achieve better workload balancing and minimal run-time data redistribution. An example is shown in Fig. 5, where a $n \times n \times n$ volume is partitioned into data blocks during pre-processing, each with the size of $m \times m \times m$. Each data block is then distributed to a unique processor later in the data distribution stage.

Due to the large data size and limited memory capacity, it may not be feasible to load all assigned data blocks into memory simultaneously at each node. To address this issue, we implement an out-of-core data management scheme similar to [5]. At each node, all the blocks assigned to it are stored in a single data file with a file description header, which keeps the necessary information such as the dimensions of the original volume, the offsets to each block in the data file, and the original block id of each block. Such information facilitates the quick retrieval of any block that is required by our run-time algorithm. In this way, only those blocks that are useful in the analysis or visualization process will be loaded into memory, thus allowing the algorithm to handle a data size beyond the memory constraint. When combined with a view-dependent algorithm, this approach further reduces the memory requirement by loading only the visible portion of the data into memory.

Campbell *et al.* [4] presented a load balancing technique for adaptive computation using space-filling curve [28] traversal, which provides us an initial insight into the workload balancing issue in parallel data analysis and visualization algorithms. Space-filling curves [28] are curves traversing the entire 2-dimensional unit square (or the 3-dimensional unit cube). In 1890, Peano discovered a densely self-intersecting curve passing through every point of the unit square, which was the first example of a space filling curve. There are many other space-filling curves available including Hilbert and Z curves. The examples of these curves in 2D space are shown in Fig. 6.

We use space-filling curves to achieve static data distribution that allows an even distribution of workload. During the data distribution process, we distribute each block to one of the processor along a space filling curve in a round-robin manner. The right plot in Fig. 5

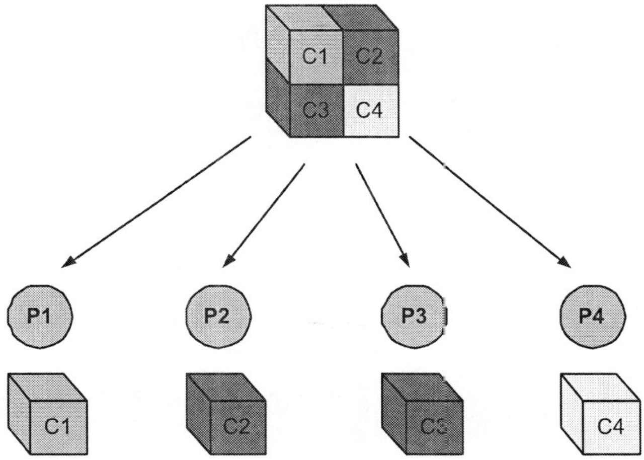

Figure 3. The volume is partitioned into chunks, each of which is assigned to one node.

shows an example of such a distribution. We summarize the reasons to use a space filling curve for data distribution as follows.

- Since the traversal along a space filling curve visits a block once and only once, by distributing the blocks along a space filling curve, each block will be assigned to only one processor, thus avoiding data duplication.
- A hierarchy of space filling curves can be built as shown in Fig. 7. The hierarchical property of a space filling curve makes it suitable to be applied to a hierarchical algorithm such as a multi-resolution algorithm, as we will show in Section 3.2.2..
- A space filling curve preserves spatial locality, i.e., the traversal path along a space-filling curve always visits the adjacent blocks before it leaves the local neighborhood. Usually a volume also has spatial coherence. For example, for a view-dependent algorithm, it is highly possible that a block is visible if its neighboring blocks are visible; for a multi-resolution algorithm, if one block satisfies a traversal threshold, it is very likely that the blocks in the neighborhood also satisfy it. Thus, if the neighboring blocks are assigned to different processors, the processors will have an equal chance to visualize blocks that are visible or satisfy a particular traversal threshold. The spatial locality in both space filling curves and data volumes helps achieve load balancing at run time.

3.2. Test Cases

We applied this data partition and distribution technique to both parallel view-dependent visualization algorithms [9, 10, 8] and parallel multi-resolution volume rendering algorithms [33, 34], which were tested on a PC cluster with 32 processing nodes. In these algorithms, data blocks are distributed statically along a space-filling curve to ensure a well-balanced workload distribution at run time, which in turn leads to high scalability. In our implementation, we tried two popular space-filling curves: a Hilbert curve and a Z curve, both of which produced qualitatively very similar results.

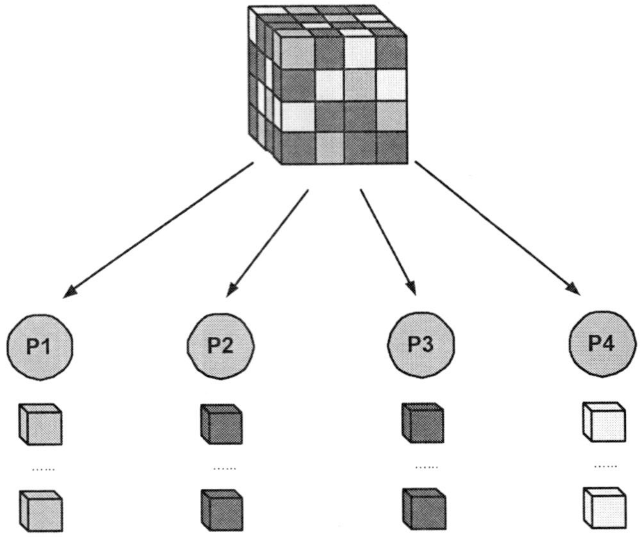

Figure 4. The entire data volume is partitioned into a number of data blocks, each of which is assigned to one node according to some distribution rules. In this example, all the blocks with the same color are assigned to the same node.

3.2.1. Parallel View-Dependent Visualization Algorithms

In Fig. 8, the small variation in the number of blocks rendered by each of the 32 processors shows that, with or without visibility culling, our algorithm can always achieve reasonably good load balancing, for two popular volume visualization algorithms, isosurface extraction and volume rendering. A high scalability is made possible by our algorithms through balanced workload. In our experiments, the view-dependent isosurface extraction algorithm achieved 88.6% and 84.5% parallel utilization on 16 and 32 processors, respectively. Similarly, the volume-rendering algorithm achieved about 89.2% and 81.4% parallel utilization on 16 and 32 processors, respectively. These results demonstrate a large improvement over

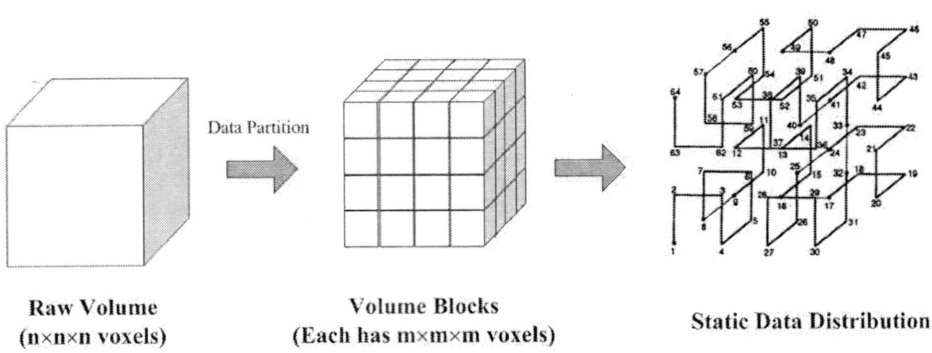

Figure 5. The process of data partition and data distribution. The number on the curve (right) represents the traversal order along a space filling curve. The corresponding blocks will be distributed to the processors in a round-robin manner following this order.

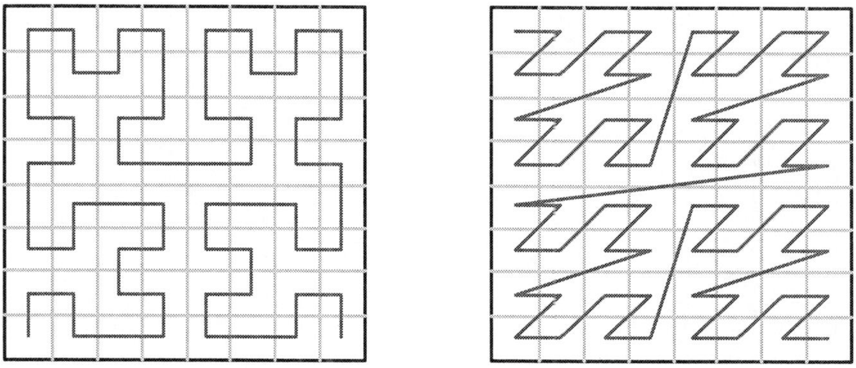

(a) a 2D Hilbert curve (b) a 2D Z curve

Figure 6. Two examples of space filling curves.

the previous algorithms [14].

We applied a similar data management and distribution scheme in our parallel visibility accelerated time-varying volume rendering algorithm [10]. At each time step, the volume is partitioned and distributed as described in Fig. 5. Our algorithm still achieved balanced workload and high scalability. Fig. 9 shows the total number of blocks rendered by each of the 32 processors for 64 time steps. During the rendering process at each time step, our parallel renderer consistently achieved higher than 80% CPU utilization on a 32-processor PC cluster.

3.2.2. Parallel Multi-Resolution Volume Rendering Algorithms

Multi-resolution rendering is a technique that visualizes data at different resolutions to balance image quality and computation speed. Many visualization algorithms provide users with the capability of multi-resolution rendering since rendering a lower resolution of the data sometimes is sufficient to obtain a quick overview of the data, especially when the resolution of the data is far beyond the highest screen resolution currently available. A multi-resolution algorithm typically builds a hierarchical data representation. Recently, Guthe *et al.* [11] presented a hierarchical wavelet representation, as shown in Fig. 10, for large data sets that support interactive walkthroughs on a single commodity PC. The root node has both highpass filtered and lowpass filtered data, while other nodes only keep highpass filtered data. Based on such a representation, we proposed a parallel multi-resolution volume rendering algorithm with an effective error-guided load balancing scheme using a hierarchical space filling curve.

To avoid data replication and run-time data redistribution, the wavelet hierarchy needs to be partitioned and distributed to processors at a preprocessing stage. However, as shown in Fig. 10, the wavelet tree structure has a strong parent-child node relationship that results in dependency among the processors if we simply distribute nodes to different processors. To resolve this issue, we proposed a *EVERY-k scheme* [33]: low resolution data blocks are reconstructed and stored for nodes at every k levels starting from the root, where $k < h$, and h is the height of the wavelet tree. We refer to a node that stores the reconstructed data block as a *representative node*, while a node that stores only wavelet coefficients as

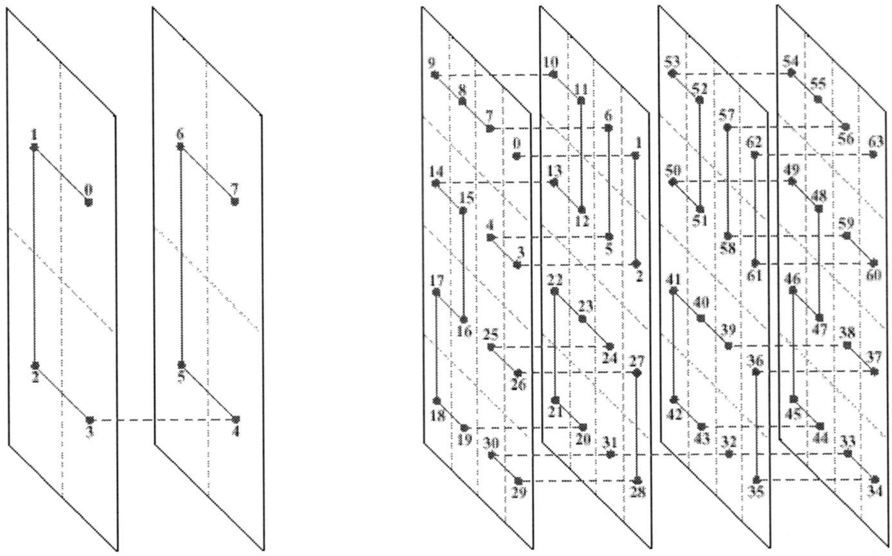

Figure 7. Two levels of 3-dimensional Hilbert curve [4], where the numbers indicate the traversal order.

an *associated node*. By default, the root of the wavelet tree is a representative node. All the leaf nodes of the tree are designated as associated nodes to reduce storage overhead. Fig. 11 shows an example where $k = 2$ and $h = 6$. In this scheme, data reconstruction only needs to be performed for the associated nodes by recursively requesting their parent nodes up to the closest representative node, where the low resolution data has already been reconstructed. We define a *distribution unit* as the data at a representative node along with the wavelet coefficients at all its descendent nodes which depend on the representative node. This definition implies that there must be one and only one representative node in a distribution unit and all the nodes in one distribution unit are independent of nodes in any other distribution units. We use the distribution units to form a *partition* of the wavelet tree, and a distribution unit is used as the *minimum* unit to be assigned to a processor. Since there is no data dependency among distribution units during wavelet reconstruction, we are able to eliminate the dependency among processors at run time.

It is challenging to design a static data distribution scheme for parallel multi-resolution algorithms that achieve good load balancing. Different data resolutions, and thus different rendering workload, could be chosen at run time for different error threshold the user may choose. However, in general, a data set usually exhibits strong spatial coherence. Given an error tolerance, if a particular data resolution is chosen for a subvolume, it is more likely that a similar resolution will also be used for the neighboring subvolumes. In our multi-resolution algorithm [33], if a block at a certain level is selected to be rendered, its neighboring blocks in the same tree level will most likely also be rendered. Thus, if neighboring data blocks at a similar resolution are evenly distributed to different processors, each processor will receive approximately the same rendering workload in that local neighborhood. Based on this consideration, a hierarchical space-filling curve was utilized in our data distribution scheme to assign the distribution units to different processors. In Fig. 12, we

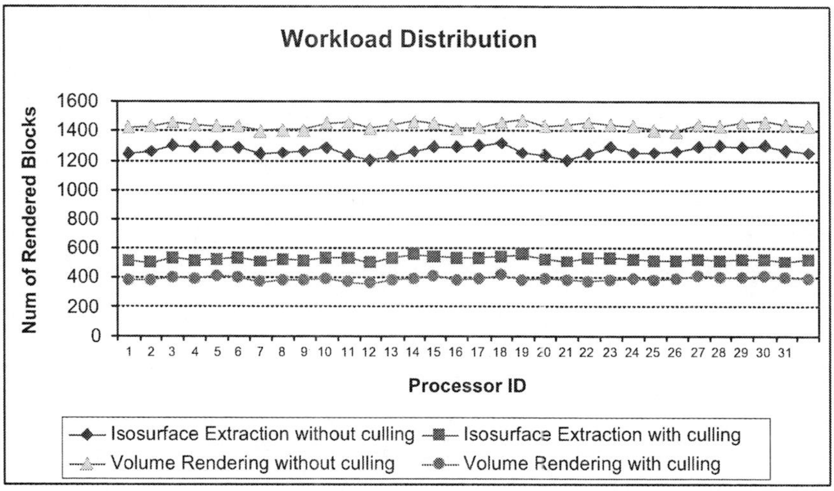

Figure 8. The number of blocks rendered with and without visibility culling for both volume rendering and isosurface extraction.

show a simplified 2D example of a multi-resolution tree and its corresponding space-filling curve at each level. The numbers in the figure show the traversal order along the hierarchal space-filling curve.

To further ensure workload balancing at run time, data blocks with similar error values should be distributed to different processors since they are most likely to be selected together for rendering. To achieve this, in addition to the hierarchal space-filling curve traversal, we include an error-guided bucketization mechanism in our data distribution scheme. As illustrated in Fig. 13, our algorithm works as follows: The entire error range $[errmin, errmax]$ is first partitioned into several error intervals, where $errmin$ and $errmax$ are the minimum and maximum error values of the representative nodes from all the distribution units. Then, we traverse along a hierarchical space-filling curve, where every distribution unit encountered is sorted, according to the traversal order, into a bucket associated with an error interval when the error value of its representative node falls into that interval range. The intervals of the buckets are adjusted so that each bucket holds a similar number of distribution units. Finally, all sorted distribution units in the buckets are distributed among processors in a round-robin manner.

The error-guided hierarchical data distribution allows our parallel multi-resolution volume rendering algorithm to effectively balance the workload among the processors. Fig. 14 shows the numbers of data blocks rendered at each of the 32 processors when three different error tolerances were used. The processors rendered approximately an equal number of blocks, resulting in a good load balancing. The well-balanced workload implies the high scalability of our parallel algorithm, which achieved approximately 97.6%, 96.8% and 87.3% parallel CPU utilization for 8, 16 and 32 processors, respectively.

We further extended our work to time-varying volume rendering. A wavelet-based time-space partitioning (WTSP) tree [34] was constructed for multi-resolution rendering of a time-varying dataset using a similar data management and distribution method. Fig. 15 shows the number of data blocks rendered at each of the 32 processors when three different

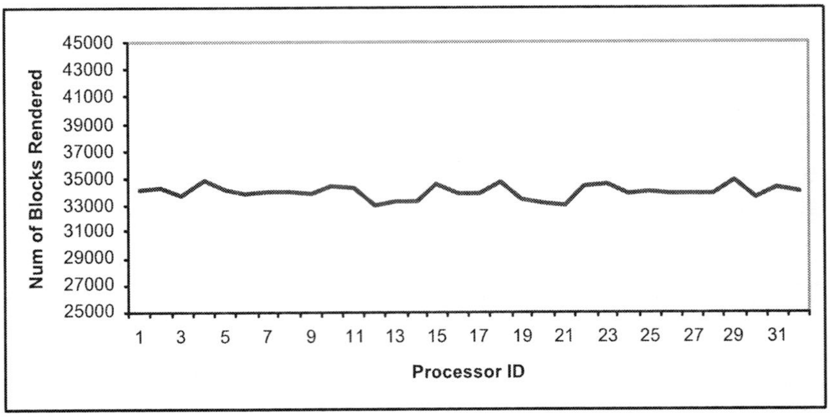

Figure 9. The number of total blocks rendered by each processor throughout 64 time steps.

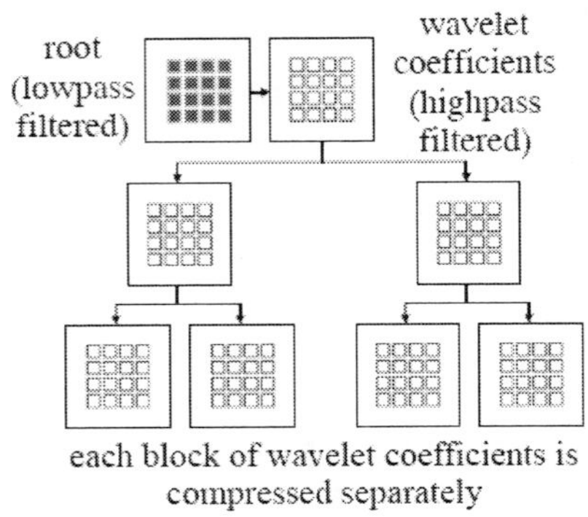

Figure 10. The compressed wavelet tree [11].

time steps and tolerances for both spatial and temporal errors were used. The well-balanced workload we achieved ensures the scalability of our parallel algorithm.

4. Image Compositing

Data partition and distribution we discussed above is usually a one-time operation. Recently, much research attention is attracted to the direction of improving run-time performance. Typically, for large-scale scientific simulation data sets, a large number of parallel processors are used to reduce data rendering time. With the increasing popularity of large screen display devices such as powerwall, high-resolution images are often generated for remote display, which significantly increases the time needed for image compositing and delivery.

Several previous research efforts have developed effective solutions to minimizing the image compositing time within the cluster. Binary swap (BS) [22] takes a minimum number of steps to perform image compositing, with a power-of-two restriction imposed on the number of processors. Parallel-pipelined (PP) algorithm [16] organizes an arbitrary number

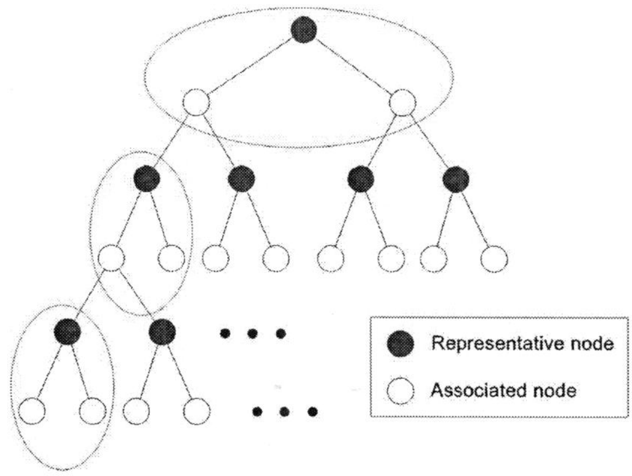

Figure 11. Only the nodes at every $k = 2$ levels (marked as shaded nodes) starting from the root store the data blocks in a tree of height $h = 6$. The ellipses show examples of distribution units. A binary tree rather than an octree is drawn here for illustration purposes.

of processor nodes in a mesh structure, which expedites the image compositing process but may produce high communication overheads if a large number of processors are used. Rotate-tiling (RT) [17] algorithm, a combination of BS and PP, employs an arbitrary number of nodes and determines the best block number to partition an image of a given size. These existing methods are mainly focused on the image compositing alone (or to be fair, their main design goal is for local rendering and local display). Since the partial final images are distributed on all participating nodes in the existing methods, a gathering step is required to generate a complete final image, which is not suitable for pipelining with image delivery over wide-area network connections.

To support remote visualization, we propose a linear pipeline (LP) scheme for image compositing and delivery. The proposed LP scheme has the following key features: (i) the processors are organized into the simplest linear structure; (ii) the algorithm works for an arbitrary number of processors; (iii) the image compositing performance is improved compared to other existing methods; and more importantly, (iv) the in-cluster image compositing and remote image delivery are pipelined so that an optimal overall system performance is achieved. In the proposed scheme, the processors within the cluster are linearly arranged, and the image is divided into a certain number of segments, which flow through the pipeline to the remote client consecutively. We analytically determine the segment size that minimizes the final image display time and derive the conditions where the proposed image compositing and delivery scheme outperforms the traditional scheme based on the BS algorithm.

4.1. Linear Pipeline for Image Compositing and Delivery

A general structure for parallel remote visualization system is illustrated in Fig. 16. A data of large size is first partitioned into a certain number of blocks that are then distributed onto the processors in the cluster. Upon successful completion of data rendering on all processors, the partial images are composited and the final image is delivered to the remote client over a wide-area network.

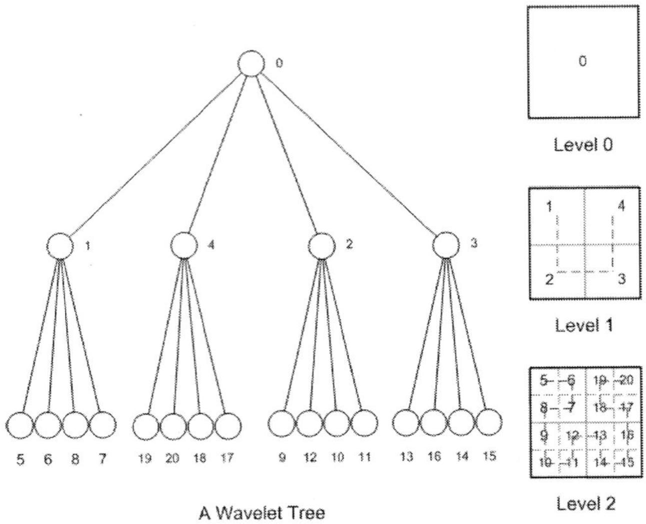

Figure 12. A simplified 2D example of data distribution along the hierarchical space-filling curve. All the multi-resolution tree nodes are traversed level by level in a breadth-first search manner. The numbers associated with the tree nodes indicate the traversal order given by the space-filling curve. A popular space-filling curve, the Hilbert curve, is used in this example.

In our linear pipeline (LP) scheme for image compositing and delivery, the image resulted from the rendering stage is partitioned into a certain number of segments on a leading processor, each of which flows through a sequence of processors. Each intermediate processor receives a segment from its immediate upstream processor, integrates the received segment with its own image segment of the same position, and sends the integrated segment to its immediate downstream processor for further integration. As shown in Fig. 17, at the end of the image compositing loop, the last processor produces a partial final image of the segment and sends it immediately over wide-area network connections to a remote client.

The advantages of data pipeline on the performance improvement of parallel image compositing for remote visualization are three fold. First, pipelining masks the processor and network latencies that are known to be an important bottleneck in high-bandwidth local-area networks. Second, it allows the simultaneous sending and receiving of data, hence exploiting the full duplex nature of the interconnection links in the parallel system. Third, pipelining image compositing and delivery makes it possible to start the delivery of partial final images even before the full final image is generated.

4.2. Mathematical Models and Performance Analysis

We consider a remote visualization system that performs data rendering and image compositing in a parallel cluster with p homogeneous processors and delivers images of size m pixels to a remote client for display over a wide-area network. Processors are able to send and receive data concurrently between each other via a single full-duplex local link

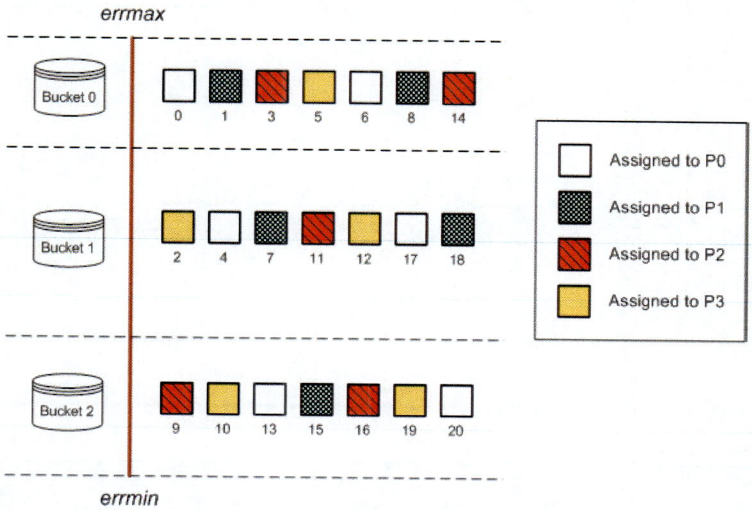

Figure 13. An example of data distribution along the hierarchical space-filling curve with error-guided bucketization. The numbers associated with the distribution units indicate the traversal order given by the space-filling curve. In this example, a total of 21 distribution units with three different resolutions are distributed among four processors.

of bandwidth B_l bits per second (bps) and delay d_l seconds [1]. The image compositing on each processor is performed at the rate of C pixels per second. The remote wide-area connection between the cluster and the client has bandwidth B_r bps and delay d_r seconds. For convenience, we introduce three new notations: $b_l = K/B_l$, $b_r = K/B_r$, $c = 1/C$, where $K = 8 \cdot n$ bits per pixel is the coefficient translating a pixel of n bytes to $8 \cdot n$ bits.

4.2.1. Image Compositing Using Binary Swap Algorithm

To facilitate the performance analysis for the binary swap (BS)-based image compositing scheme, we assume that the number of processors p is a power of two. Note that this assumption is eliminated in the proposed linear pipeline scheme. We consider three separated steps: binary swap, image gathering, and image delivery. The time $t_{BS}(BinarySwap)$ for binary swap is defined as:

$$t_{BS}(BinarySwap) = \sum_{i=1}^{\log_2^p} \left((b_l + c) \frac{m}{2^i} + d_l \right). \quad (1)$$

The time $t_{BS}(ImageGathering)$ for image gathering is defined as:

$$t_{BS}(ImageGathering) = \sum_{i=1}^{\log_2^p} \left(b_l \frac{m}{2^i} + d_l \right). \quad (2)$$

[1]The delay d denotes the minimum link delay consisting of all non-bandwidth-constrained delay elements such as link propagation delay and queuing delay.

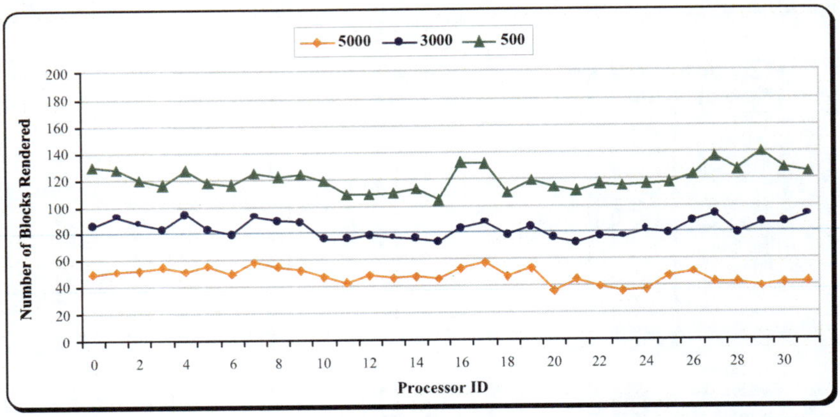

Figure 14. The number of data blocks rendered at each of the 32 processors with three different error tolerances. A total of 1,510, 2,640, and 3,850 blocks were rendered for error tolerances of 5,000, 3,000, and 500, respectively.

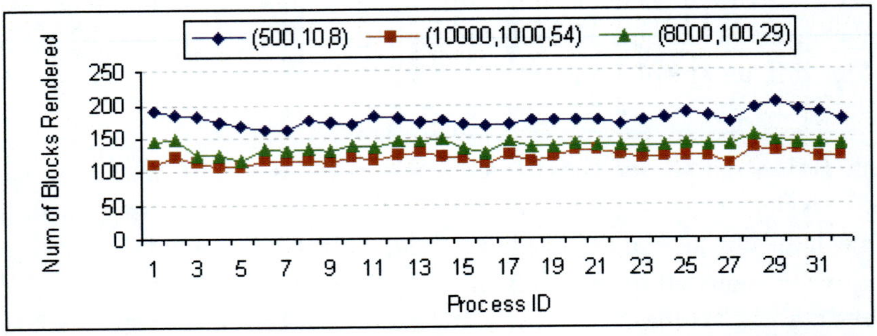

Figure 15. The number of data blocks rendered at each of the 32 processors with three different queries of (se, te, t) where se is the spatial error tolerance, te is the temporal error tolerance and t is the time step. A total of 5690, 3802, and 4379 blocks were rendered for $(500, 10, 8)$, $(10000, 1000, 54)$ and $(8000, 100, 29)$, respectively.

The time $t_{BS}(ImageDelivery)$ for delivering the final image is defined as:

$$t_{BS}(ImageDelivery) = mb_r + d_r. \tag{3}$$

The total time T_{BS} for binary swap, image gathering, and image delivery is defined as:

$$\begin{aligned} T_{BS} &= T_{BS}(BinarySwap) + T_{BS}(ImageGathering) + T_{BS}(ImageDelivery) \\ &= \sum_{i=1}^{\log_2^p} \left((b_l + c)\frac{m}{2^i} + d_l \right) + \sum_{i=1}^{\log_2^p} \left(b_l \frac{m}{2^i} + d_l \right) + mb_r + d_r \\ &= (2d_l \log_2^p + d_r) + \left((2b_l + c)(1 - \frac{1}{p}) + b_r \right) m. \end{aligned} \tag{4}$$

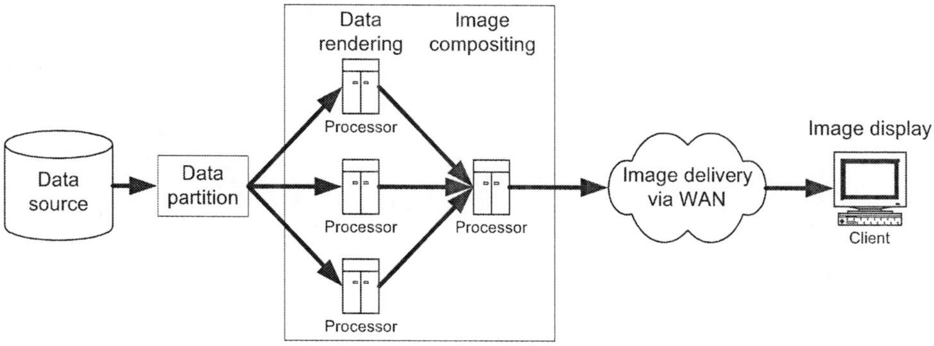

Figure 16. The structure of a general parallel-based remote visualization system.

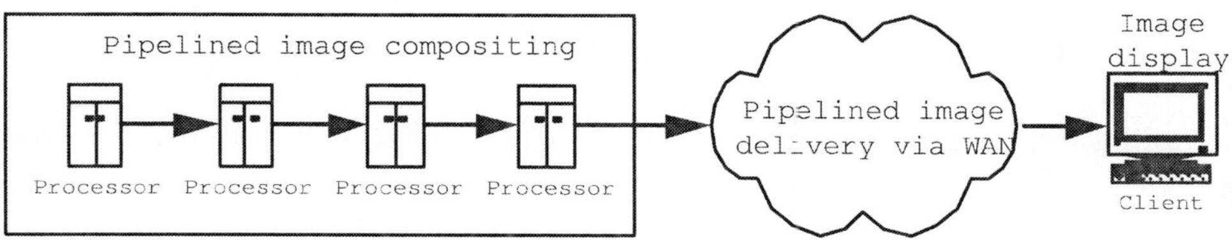

Figure 17. Pipelined image compositing and delivery.

4.2.2. Image Compositing Using Linear Pipeline Algorithm

In the linear pipeline image compositing scheme, all the processors in the parallel system and the remote client are linearly arranged to form a chain. The image of m pixels is partitioned into t segments of s pixels, i.e. $m = ts$. Every segment sequentially flows through the chain for image compositing and delivery. We denote the j-th image segment on the i-th processor as $m[i][j]$. The task of image compositing is to calculate the image data as $m[0][j]$ *op* $m[1][j]$ *op* ... *op* $m[p-1][j]$ for $0 \leq j \leq t-1$ in a pipelining manner and deliver the completed image segments to the remote client sequentially as soon as they are available. Note that here we use "op" to denote the image compositing operation.

In Fig. 18, we illustrate the proposed linear pipeline image compositing scheme for remote visualization using a cluster of four processors. In step 0, processor 0 sends image segment $m[0][0]$ to processor 1. Processor 1 receives this segment from processor 0 and calculate $m[0][0]$ *op* $m[1][0]$. In step 1, processor 0 sends image segment $m[0][1]$ to processor 1. Processor 1 receives this new segment and sends partially completed image segment $m[0][0]$ *op* $m[1][0]$ to processor 2 concurrently, and then calculates $m[0][1]$ *op* $m[1][1]$. Processor 2 first receives image segment $m[0][0]$ *op* $m[1][0]$ from processor 1 and then calculates $m[0][0]$ *op* $m[1][0]$ *op* $m[2][0]$. As the procedure continues, in step 2, processor 3 (the last one in the cluster) has the first completed image segment $m[0][0]$ *op* $m[1][0]$ *op* $m[2][0]$ *op* $m[3][0]$. In step 3 and onwards, the remote client receives a completed image segment at the end of each step. A final full image is displayed when all image segments arrive at the remote client.

In the linear pipeline image compositing scheme, we spend the first $p-1$ steps establishing the pipeline, each of which takes $(b_l + c)s + d_l$. The time $T_{LP}(EstablishingPipeline)$

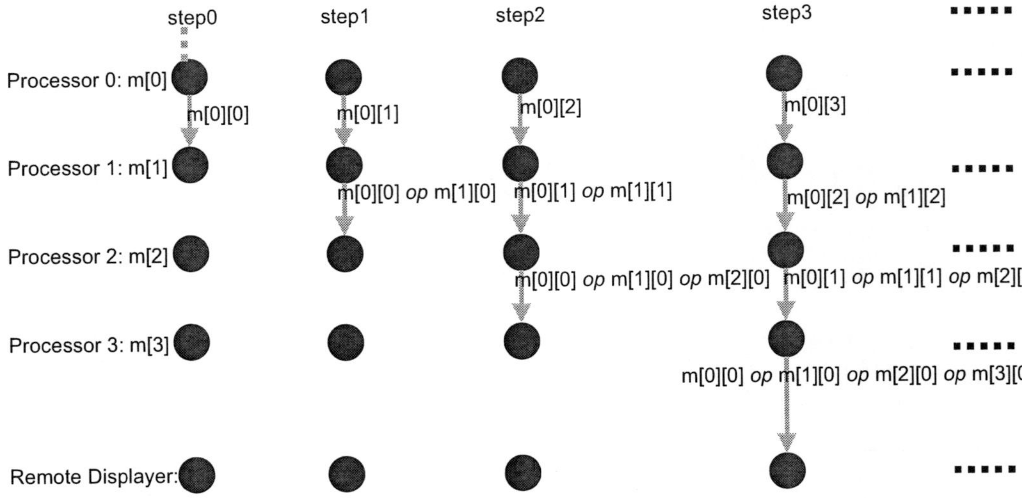

Figure 18. Illustration of linear pipeline image compositing for remote visualization on four processors.

for pipeline establishment is defined as:

$$T_{LP}(EstablishingPipeline) = (p-1)\left((b_l + c)s + d_l\right). \tag{5}$$

Once the pipeline is established, the time of each step to deliver a completed image segment to the remote client is determined by the pipeline bottleneck, i.e. the maximum of the in-cluster image segment generation rate and remote data transport speed over the wide-area network. Since there are t segments in total, we define the time $T_LP(RestSteps)$ for the rest steps as:

$$T_{LP}(RestSteps) = t \cdot \max\{(b_l + c)s + d_l, b_r s + d_r\}. \tag{6}$$

Considering two cases where the pipeline bottleneck is located either within the cluster or on the remote connection, the total time for image compositing and delivery is calculated as follows:

$$\begin{aligned} T_{LP}(s) &= T_{LP}(EstablishingPipeline) + T_{LP}(RestSteps) \\ &= (p-1)\left((b_l + c)s + d_l\right) + t \cdot \max\{(b_l + c)s + d_l, b_r s + d_r\} \\ &= \begin{cases} (p-1)(b_l + c)s + (p-1)d_l + (b_l + c)m + \frac{md_l}{s}, \\ \qquad\qquad\qquad \text{if } (b_l + c)s + d_l > b_r s + d_r. \\ \\ (p-1)(b_l + c)s + (p-1)d_l + b_r m + \frac{md_r}{s}, \\ \qquad\qquad\qquad \text{if } (b_l + c)s + d_l < b_r s + d_r. \end{cases} \end{aligned} \tag{7}$$

In the above two cases where $(b_l + c)s + d_l \neq b_r s + d_r$, since $T_{LP}(s)$ has a nonnegative second order derivative:

$$T''_{LP}(s) \geq 0,$$

and

$$\lim_{s \longrightarrow \infty} T_{LP}(s) = \lim_{s \longrightarrow 0} T_{LP}(s) = \infty,$$

there exists a minimum of $T_{LP}(s)$ at a certain s value, which satisfies the condition:

$$T'_{LP}(s) = 0.$$

The bottleneck location of the image compositing and delivery pipeline must be in one of the following three cases, including the one where $(b_l + c)s + d_l = b_r s + d_r$.

- Case 1: if $(b_l + c)s + d_l > b_r s + d_r$, then

$$T_{LP}(s) = (p-1)(b_l + c)s + (p-1)d_l + (b_l + c)m + \frac{md_l}{s}. \quad (8)$$

  We calculate the first-order derivative of $T_{LP}(s)$ and set it to zero:

$$T'_{LP}(s) = -\frac{md_l}{s^2} + (p-1)(b_l + c) = 0.$$

  By solving the above equation, we obtain the first possible value of s that makes $T_{LP}(s)$ reach its minimum:

$$s_1 = \sqrt{\frac{md_l}{(p-1)(b_l + c)}}. \quad (9)$$

- Case 2: if $(b_l + c)s + d_l < b_r s + d_r$, then

$$T_{LP}(s) = (p-1)(b_l + c)s + (p-1)d_l + b_r m + \frac{md_r}{s}.$$

  Similarly, we caculate the first-order derivative of $T_{LP}(s)$ and set it to zero:

$$T'_{LP}(s) = -\frac{md_r}{s^2} + (p-1)(b_l + c) = 0.$$

  By solving the above equation, we obtain the second possible value of s that makes $T_{LP}(s)$ reach its minimum:

$$s_2 = \sqrt{\frac{md_r}{(p-1)(b_l + c)}}, \quad (10)$$

- Case 3: if $(b_l + c)s + d_l = b_r s + d_r$, then we have the third possible value of s that makes $T_{LP}(s)$ reach its minimum:

$$s_3 = \frac{d_r - d_l}{b_l + c - b_r}. \quad (11)$$

If $T_{LP}(s)$ reaches its minimum at $s_1 = \sqrt{\frac{md_l}{(p-1)(b_l+c)}}$, then

$$T_{LP}^{min} = (p-1)d_l + (b_l+c)m + 2\sqrt{(p-1)d_l(b_l+c)m}$$
$$= (p-1)d_l + (b_l+c)m\left(1 + 2\sqrt{\frac{d_l}{b_l+c}}\sqrt{\frac{p-1}{m}}\right). \quad (12)$$

In practice, we have a large image size m and a relatively small number p of processors, typically resulting in a negligible quantity $2\sqrt{\frac{(p-1)d_l}{b_l+c}}\sqrt{\frac{1}{m}}$. Hence, the minimum time can be approximated by:

$$T_{LP}^{min}(Case\ 1) \approx (p-1)d_l + (b_l+c)m. \quad (13)$$

If $T_{LP}(s)$ reaches its minimum at $s_2 = \sqrt{\frac{md_r}{(p-1)(b_l+c)}}$, then

$$T_{LP} = (p-1)d_l + b_r m + 2\sqrt{(p-1)d_r(b_l+c)m}$$
$$= (p-1)d_l + b_r m\left(1 + 2\sqrt{\frac{d_r(b_l+c)}{b_r^2}}\sqrt{\frac{p-1}{m}}\right). \quad (14)$$

For a large image size m and a small number p of processors, the quantity $2\frac{\sqrt{(p-1)d_r(b_l+c)}}{b_r}\sqrt{\frac{1}{m}}$ is negligible. We obtain an approximate minimum of T_{LP} as:

$$T_{LP}^{min}(Case\ 2) \approx (p-1)d_l + b_r m. \quad (15)$$

If $T_{LP}(s)$ reaches its minimum at $s_3 = \frac{d_r-d_l}{b_l+c-b_r}$, then from the concave property of $T_{LP}(s)$, we have:

$$min\{s_1, s_2\} \le s_3 \le max\{s_1, s_2\}.$$

It follows that

$$s_3 \le max\{s_1, s_2\}$$
$$= \sqrt{\frac{max\{d_l, d_r\}}{b_l+c}}\sqrt{\frac{m}{p-1}} \quad (16)$$

and

$$\frac{1}{s_3} \le \frac{1}{min\{s_1, s_2\}}$$
$$= \sqrt{\frac{b_l+c}{min\{d_l, d_r\}}}\sqrt{\frac{p-1}{m}}. \quad (17)$$

Plugging the above two inequalities into Eq. 8, we have

$$\begin{aligned}
T_{LP} &= (p-1)d_l + (b_l+c)m + \frac{md_l}{s_3} + (p-1)(b_l+c)s_3 \\
&\leq (p-1)d_l + (b_l+c)m + md_l \cdot \sqrt{\frac{b_l+c}{min\{d_l,d_r\}}}\sqrt{\frac{p-1}{m}} \\
&\quad +(p-1)(b_l+c)\cdot\sqrt{\frac{max\{d_l,d_r\}}{b_l+c}}\sqrt{\frac{m}{p-1}} \\
&= (p-1)d_l + (b_l+c)m + \left(d_l\sqrt{\frac{b_l+c}{min\{d_l,d_r\}}} + \sqrt{max\{d_l,d_r\}(b_l+c)}\right)\sqrt{(p-1)m} \\
&= (p-1)d_l + (b_l+c)m\left(1+\sqrt{\frac{1}{b_l+c}}\left(\sqrt{\frac{d_l^2}{min\{d_l,d_r\}}} + \sqrt{max\{d_l,d_r\}}\right)\sqrt{\frac{p-1}{m}}\right) \quad (18)
\end{aligned}$$

Let

$$W = max\left\{2\sqrt{\frac{d_l}{b_l+c}}, 2\sqrt{\frac{d_r(b_l+c)}{b_r^2}}, \sqrt{\frac{1}{b_l+c}}\left(\sqrt{\frac{d_l^2}{min\{d_l,d_r\}}} + \sqrt{max\{d_l,d_r\}}\right)\right\}.$$

Therefore, in either one of the three cases where $T_{LP}(s)$ reaches its minimum at s_1, s_2, or s_3, we have

$$T_{LP} \leq (p-1)d_l + max\{b_l+c,\ b_r\}\cdot m\cdot\left(1+W\cdot\left(\sqrt{\frac{p-1}{m}}\right)\right). \quad (19)$$

In the current computing environments for large-scale visualization, we typically have an image size m on the order of dozens or hundreds of MBytes and a relatively small number of processors less than one thousand, which results in a small $W\sqrt{(p-1)/m}$ much less than one. Therefore, we have

$$T_{LP} \approx (p-1)d_l + max\{b_l+c\ ,\ b_r\}m. \quad (20)$$

4.2.3. Analytical Performance Comparison

To make the performance comparison feasible between the binary swap and linear pipeline schemes, we assume that the number p of processors is a power of two. We now derive the conditions where the linear pipeline scheme performs better or worse than the binary swap scheme. Again, we consider two cases where the pipeline bottleneck is located either within the cluster or over the remote transport.

- Case 1: when $W\sqrt{(p-1)/m}$ is small, we have

$$\begin{aligned}
T_{LP} &\leq (p-1)d_l + max\{b_l+c\ ,\ b_r\}\cdot m\cdot\left(1+W\cdot\left(\sqrt{\frac{p-1}{m}}\right)\right) \\
&\approx max\{b_l+c\ ,\ b_r\}\cdot m \quad (21)
\end{aligned}$$

while

$$T_{BS} = (2d_l \log_2 p + d_r) + \left((2b_l + c)(1 - \frac{1}{p}) + b_r\right) m$$
$$\approx ((2b_l + c) + b_r) \cdot m \quad (22)$$

The linear pipeline scheme achieves a better performance than the binary swap scheme

- Case 2: when $W\sqrt{(p-1)/m}$ is large, the linear pipeline scheme achieves a worse performance than the binary swap scheme.

As indicated above, in today's parallel processing environment, the number of processors is usually on the order of hundreds or thousands. Since the images are typically of MBytes, $W\sqrt{(p-1)/m}$ is often small. Therefore, in many cases, the linear pipeline scheme achieves a much better performance than the binary swap scheme. We will illustrate the performance superiority of the proposed LP scheme in the section of experimental results.

4.3. Implementation and Results

The image compositing schemes including both binary swap (BS) and linear pipeline (LP) are implemented under Linux in C language. We perform image compositing on the hawk PC cluster at Oak Ridge National Laboratory (ORNL) and deliver the final images to a remote computer.

4.3.1. Optimal Segment Size in LP

In order to justify the correctness of our derivation for the optimal segment size in the LP scheme, we apply the LP image compositing to an image of $m = 11520 \times 3072$ pixels[2] using $p = 16$ processors. The network transport tool [36] measures the local bandwidth within the hawk cluster B_l = 1771.54 Mbps with a minimum node-to-node delay d_l = 59 microseconds, and the remote bandwidth over wide-area connections B_r = 554.0 Mbps with a minimum end-to-end delay d_r = 209 microseconds. We also estimate that each cluster node processes about C=13583915 pixels per second for image compositing operations. The performance measurements in response to various segment sizes are shown in Fig. 19, where x axis represents the logarithm of the segment sizes in pixels and y axis represents the measured time in seconds for LP image compositing. The optimal segment size is theoretically calculated by Eq. 9 in Case 1 to be 38965 pixels producing a minimum image compositing and delivery time of 3.3524 seconds, which are consistent with the corresponding measurements in the experiments as shown in Fig. 19.

4.3.2. Performance Comparison Between BS and LP

Using the similar network and computing configurations in Subsection 4.3.1., we theoretically calculate and measure through experiments the image compositing time of BS and

[2]This corresponds to the resolution of the powerwall display at Oak Ridge National Laboratory.

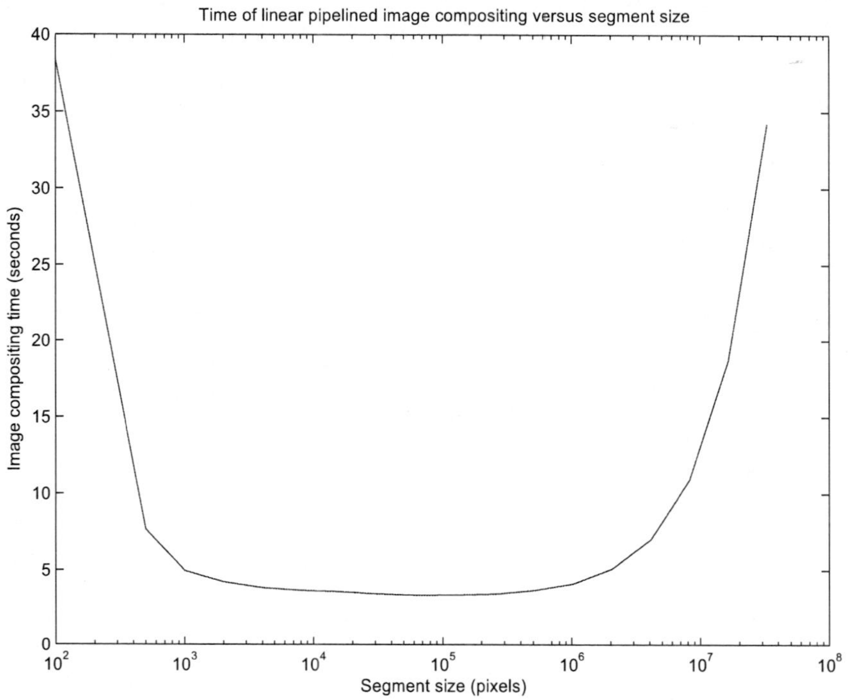

Figure 19. Image compositing time versus segment size using LP scheme.

LP in response to the number of processors and image size. The theoretical and experimental results as well as the performance comparisons are tabulated in Fig. 20, where TBS denotes the theoretically calculated (or predicted) time in seconds for BS image compositing, MBS denotes the measured time in seconds for BS image compositing, $EBS = |TBS - MBS|/MBS$ denotes the absolute relative error in percentage between predicted and measured time for BS image compositing; TLP, MLP, and ELP have the similar meanings for LP image compositing; $(MBS - MLP)/MLP$ denotes the relative speedup achieved by LP over BS (a positive sign indicates performance improvement while a negative sign indicates performance degradation). As we observe in Fig. 20, it generally takes a longer time to composite partial images from a larger number of processors [3]. Except for small image sizes, the theoretical computing models we derived for BS and LP produce accurate performance predictions. In most cases, the LP scheme is able to achieve $40\% \sim 70\%$ performance speedup over the BS scheme. Consistent with our derivation for the condition $W\sqrt{(p-1)/m}$, we also noticed that a large number of processors p and a small image size m may violate this condition.

For the convenience of performance comparison, we plot the numerical results in Fig. 20 in Fig. 21, where the top window depicts the measured 3-D performance profiles in terms of image compositing time in seconds while the bottom one depicts the theoretically calculated performance profiles using both BS (upper surface) and LP (lower surface) schemes.

[3]This observation reflects the tradeoff between an increment in image compositing time and a decrement in image rendering time when the number of processors increases.

Image size (pixels)	Number of processors													
	4							8						
	TBS (s)	MBS (s)	EBS (%)	TLP (s)	MLP (s)	ELP (%)	$\frac{MBS-MLP}{MLP}$ (%)	TBS (s)	MBS (s)	EBS (%)	TLP (s)	MLP (s)	ELP (%)	$\frac{MBS-MLP}{MLP}$ (%)
17280	.003	.011	73.9	.003	.010	68.3	10.0	.003	.013	75.2	.004	.013	65.7	0.0
34560	.005	.016	66.9	.005	.012	58.4	33.3	.006	.016	63.3	.007	.015	55.2	6.7
69120	.010	.018	43.7	.008	.016	49.0	12.5	.011	.022	49.1	.010	.019	44.7	15.8
138240	.020	.030	33.9	.016	.021	24.5	42.9	.022	.032	31.8	.017	.025	32.0	28.0
276480	.039	.051	23.2	.029	.033	9.8	54.5	.043	.056	23.1	.032	.036	10.5	55.6
552960	.078	.096	18.8	.057	.060	5.2	60.0	.086	.105	18.5	.060	.062	2.8	69.4
1105920	.155	.185	16.0	.110	.120	8.3	54.2	.171	.201	15.1	.115	.128	10.3	57.0
2211840	.310	.358	13.3	.215	.225	4.4	59.1	.341	.393	13.3	.221	.236	6.1	66.5
4423680	.620	.674	8.0	.423	.449	5.8	50.1	.681	.660	3.2	.432	.460	6.1	43.5
8847360	1.24	1.21	2.7	.835	.842	0.8	43.3	1.36	1.34	1.7	.848	.867	2.2	54.4
17694720	2.48	2.45	1.1	1.66	1.66	0.3	47.5	2.72	2.61	4.3	1.67	2.02	17.1	29.2
35389440	4.96	4.85	2.2	3.29	3.24	1.6	49.7	5.44	5.23	4.1	3.32	3.50	5.3	49.3

Image size (pixels)	Number of processors													
	16							32						
	TBS (s)	MBS (s)	EBS (%)	TLP (s)	MLP (s)	ELP (%)	$\frac{MBS-MLP}{MLP}$ (%)	TBS (s)	MBS (s)	EBS (%)	TLP (s)	MLP (s)	ELP (%)	$\frac{MBS-MLP}{MLP}$ (%)
17280	.003	.016	78.4	.006	.024	73.6	-33	.004	.021	82.7	.009	.069	86.6	-69
34560	.006	.019	67.2	.009	.025	63.3	-24	.006	.022	70.6	.013	.072	82.1	-69
69120	.012	.026	54.7	.014	.029	52.4	-10	.012	.028	56.6	.019	.078	76.1	-64
138240	.023	.036	36.4	.021	.036	40.3	0	.023	.040	41.3	.028	.086	67.5	-53
276480	.045	.059	23.6	.035	.049	29.2	20.4	.046	.066	30.1	.043	.104	58.2	-36
552960	.090	.109	17.9	.065	.078	16.7	39.7	.092	.115	20.4	.070	.130	46.1	-11
1105920	.178	.192	7.1	.121	.156	22.3	23.1	.182	.204	10.7	.130	.210	37.9	-2.8
2211840	.356	.409	13.0	.230	.268	14.0	52.6	.364	.423	14.0	.243	.415	41.4	1.9
4423680	.711	.780	8.8	.444	.513	13.4	52.1	.727	.813	10.6	.462	.645	28.4	26.1
8847360	1.42	1.46	2.7	.866	1.06	18.3	37.8	1.45	1.53	4.7	.890	1.07	17.3	41.6
17694720	2.84	2.77	2.7	1.70	1.96	13.5	40.9	2.91	2.91	0.2	1.73	2.01	14.0	44.4
35389440	5.69	5.42	4.9	3.35	3.55	5.7	52.4	5.81	5.74	1.2	3.40	3.65	6.8	57.2

Figure 20. Image compositing performance predictions and measurements.

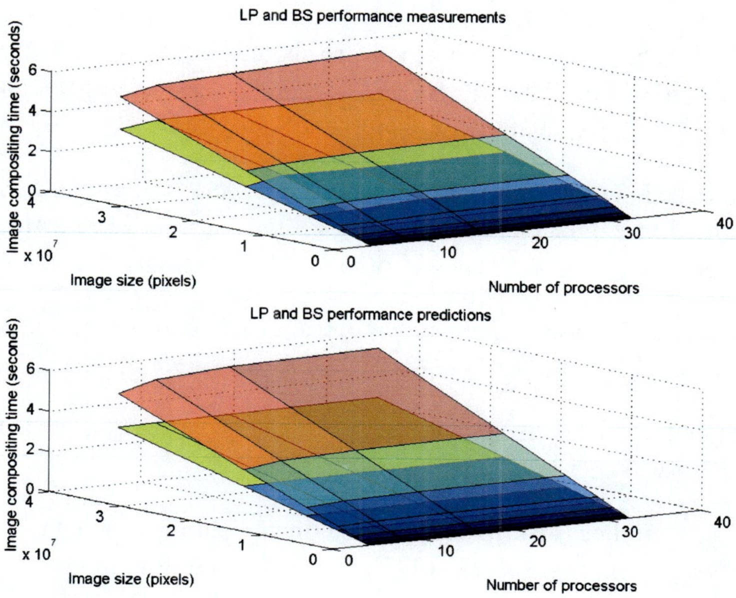

Figure 21. Image compositing time versus number of processors and segment size using LP (lower surface) and BS (upper surface). Top plot: performance measurements, bottom plot: performance predictions.

From Fig. 21, we observed that the performance of LP is superior to that of BS as long as $W\sqrt{(p-1)/m}$ is small and the performance advantage becomes more evident when the image size increases. It is worth pointing out that the similarity between the top and bottom plots in Fig. 21 also justifies the correctness of our derivations for the image compositing time using LP scheme (i.e. Eq. 20) or BS scheme (i.e. Eq. 22).

5. Conclusion

In this chapter, we present several novel techniques that target the improvement of scalability and throughput of parallel algorithms for large scale visualization by balancing workload and minimizing communication cost. We first partition the volume into a number of data blocks for flexible workload distribution, and then assign the data blocks to processors in a round-robin manner along a space-filling curve. A linear pipeline algorithm is used to speed up image compositing after a partial image is rendered at each participating processor. The effectiveness of these solutions are either mathematically proven or demonstrated using experimental data in practical applications of large-scale data analysis and visualization.

References

[1] Terascale Supernova Initiative (TSI). http://www.phy.ornl.gov/tsi.

[2] National Leadership Computing Facility (NLCF). http://www.ccs.ornl.gov/nlcf.

[3] W. Bethel, B. Tierney, J. lee, D. Gunter, and S. Lau. Using high-speed wans and network data caches to enable remote and distributed visualization. In *Supercomputing '00: Proceedings of the 2000 ACM/IEEE conference on Supercomputing (CDROM)*, Washington, DC, USA, 2000. IEEE Computer Society.

[4] P.C. Campbell, K.D. Devine, J.E. Flaherty, L.G. Gervasio, and J.D. Teresco. Dynamic Octree Load Balancing Using Space-Filling Curves. Technical Report CS-03-01, Williams College Department of Computer Science, 2003.

[5] M. Cox and D. Ellsworth. Application-controlled demand paging for out-of-core visualization. In *IEEE Visualization'97*, pages 235–244, 1997.

[6] P. Crossno and E. Angel. Isosurface extraction using particle systems. In *IEEE Visualization'97*, pages 495–498, 1997.

[7] P. Ellsiepen. Parallel isosurfacing in large unstructed datasets. In *Visualization in Scientifc Computing*, pages 9–23, 1995.

[8] J. Gao, J. Huang, C. R. Johnson, S. Atchley, and J. A. Kohl. Distributed Data Management for Large Volume Visualization. In *IEEE Visualization '05*, pages 183–189, 2005.

[9] J. Gao, J. Huang, H.-W. Shen, and J. A. Kohl. Visibility Culling Using Plenoptic Opacity Functions for Large Data Visualization. In *IEEE Visualization '03*, pages 341–348, 2003.

[10] J. Gao, H.-W. Shen, J. Huang, and J. A. Kohl. Visibility Culling for Time-Varying Volume Rendering Using Temporal Occlusion Coherence. In *IEEE Visualization '04*, pages 147–154, 2004.

[11] S. Guthe, M Wand, J. Gonser, and StraBer W. Interactive rendering of large volume data sets. In *IEEE Visualization'02*, 2002.

[12] R. Haimes. Unsteady Visualization of Grand Challenge Size CFD Problems: Traditional Post-Processing vs. Co-Processing. In *ICASE/LaRc Symposium on Visualizing Time-Varying Data*, pages 63–75, 1996.

[13] C. Hansen and P. Hinker. Massively parallel isosurface extraction. In *IEEE Visualization 1992*, 1992.

[14] J. Huang, N. Shareef, R. Crawfis, P. Sadayappan, and K. Mueller. A parallel splatting algorithm with occlusion culling. In *3rd Eurographics Workshop on Parallel Graphics and Visualization*, pages 125–132. Girona, Spain, 2000.

[15] A. Hutanu, G. Allen, S.D. Beck, P. Holub, H. Kaiser, A. Kulshrestha, M. Lika, J. MacLaren, L. Matyska, R. Paruchuri, S. Prohaska, E. Seidel, B. Ullmer, and S. Venkataraman. Distributed and collaborative visualization of large data sets using high-speed networks. *Future Generation Computer Systems*, 22(8):1004–1010, 2006.

[16] T.-Y. Lee. Image composition schemes for sort-last polygon rendering on 2d mesh multicomputers. *IEEE Transactions on Visualization and Computer Graphics*, 2(3):202–217, 1996.

[17] C.F. Lin, S.K. Liao, and Y.C. Chung. A rotate-tiling image compositing method for sort-last parallel volume rendering systems on distributed memory multicomputers. *Journal of Information Science and Engineering*, pages 643–664, 2004.

[18] E. J. Luke and C. D. Hansen. Semotus visum: a flexible remote visualization framework. In *Proc. of IEEE Visualization*, pages 61–68, 2002.

[19] E. Lum, K. Ma, and J. Clyne. A hardware-assisted scalable solution for interactive volume rendering of time-varying data. *IEEE Transactions on Visualization and Computer Graphics*, 8(3):286–301, 2002.

[20] K.-L. Ma. Runtime Volume Visualization for Parallel CFD. In *Proc. of Parallel CFD'95*, 1995.

[21] K.-L. Ma and D. M. Camp. High performance visualization of time-varying volume data over a wide-area network status. In *Proceedings of Supercomputing Conference '00*, pages 59–59, 2000.

[22] K.-L. Ma, J. S. Painter, C. D. Hansen, and M. F. Krogh. Parallel volume rendering using binary-swap compositing. *IEEE Computer Graphics and Applications*, 14(4):59–68, 1994.

[23] A. et al (Ed.) Mezzacappa. In *Proc. of SciDAC 2005: Scientific Discovery through Advanced Computing*. Institute of Physics Publishing, 2005.

[24] S. Molnar, M. Cox, D. Ellsworth, and H. Fuchs. A sorting classification of parallel rendering. *IEEE Computer Graphics and Applications*, 14(4):23–32, 1994.

[25] R. Mount. The Office of Science Data-Management Challenge. Technical Report Technical Report SLAC-R-782, Stanford Linear Accelerator Center, 2004.

[26] S. G. Parker and C. R. Johnson. Scirun: A scientific programming environment for computational steering. In *ACM/IEEE Supercomputing'95*, 1995.

[27] J. Rowlan, E. Lent, N. Gokhale, and S. Bradshaw. A Distributed, Parallel, Interactive Volume Rendering Package. In *IEEE Visualization '94*, pages 21–30, 1994.

[28] H. Sagan. *Space-filling Curves*. New York, 1994.

[29] R. Samanta, T. Funkhouser, and K. Li. Parallel rendering with k-way replication. In *IEEE 2001 Symposium on Parallel and Large-data Visualization and Graphics*, pages 75–84. IEEE Press, 2001.

[30] R. Samanta, T. Funkhouser, K. Li, and J.P. Singh. Hybrid sort-first and sort-last parallel rendering with a cluster of pcs. In *Eurographics Woskshop on Graphics Hardware*. ACM SIGGRAPH, August 2000.

[31] H.-W. Shen, C. D. Hansen, Y. Livnat, and C. R. Johnson. Isosurfacing in span space with utmost efficiency (ISSUE). In *IEEE Visualization '96*, pages 287–294, 1996.

[32] J. Simpson, A. Sanderson, E. Luke, K. Balling, and R. Coffey. Collaborative remote visualization. In *Proceedings of the IEEE/ACM SC2002 Conference*, November 2002.

[33] C. Wang, J. Gao, and H.-W. Shen. Parallel Multiresolution Volume Rendering of Large Data Sets with Error-Guided Load Balancing. In *Accepted and to appear in 2004 Eurographics Symposium on Parallel Graphics and Visualization*, 2004.

[34] C. Wang, L. Li, J. Gao, and H.-W. Shen. A Multiresolution Data Management Framework for Large Scale Time-Varying Volume Rendering. In *Submitted to IEEE Visualization 2004 conference.*

[35] Q. Wu, J. Gao, M. Zhu, N.S.V. Rao, J. Huang, and S.S. Iyengar. On optimal resource utilization for distributed remote visualization. *Accepted by IEEE Transactions on Computers*, 2007.

[36] Q. Wu, N.S.V. Rao, and S.S. Iyengar. On transport daemons for small collaborative applications over wide-area networks. In *Proc. of the 24th IEEE International Performance Computing and Communications Conference*, Phoenix, Arizona, April 7-9 2005.

[37] X. Zhang, C. Bajaj, and V. Ramachandran. Parallel and out-of-core view dependent isocontour visualization using random data distribution. In *Proc. of the Symposium on Data Visualization 2002*, 2002.

In: Cluster Computing and Multi-Hop Network Research ISBN 978-1-60876-186-9
Editors: C. Jimenez and M. Ortego, pp. 137-161

Chapter 7

Mobility-Induced Capacity-Delay Tradeoffs in Wireless Multihop Networks

Vasileios Karyotis* and Symeon Papavassiliou†
Network Management & Optimal Design Lab (NETMODE)
School of Electrical and Computer Engineering
National Technical University of Athens (NTUA)

Abstract

Wireless mobile multihop networks, such as ad hoc and sensor networks, have become rather popular in realizing seamless distributed computation and accommodating autonomic behavior. In these networks, the lack of central infrastructure, the spatio-temporal dynamic nature of wireless communications along with their mobility-induced dynamic topology, pose additional challenges on the traffic-carrying capacity of the system and raise concerns on the delay performance of packet delivery mechanisms. Despite such impediments of the wireless multihop environment, it has been recently shown that mobility can be exploited by various relay techniques for increasing the capacity of such networks. However, the throughput-effectiveness of the mobility-exploiting relay methods takes place at the cost of a greater number of intermediate hops traversed by packets traveling in the network in order to reach their final destination and additional buffering required at the intermediate nodes, revealing an inherent tradeoff between capacity and delay in mobile multihop networks. In this chapter, we first briefly review various mobility models, and then focus on the inherent mobility-induced tradeoff between capacity and delay in multihop networks. Mobility is treated as both a constructive (increasing capacity), as well as a destructive (increasing delay) factor in wireless infrastructureless networks. Formal problem settings are presented, followed by applications and protocols that could be used in actual networks for increasing performance and realizing viable networking services. We also provide a qualitative and quantitative comparison among the various protocols, from which directions for future research studies are indicated.

Key Words: capacity, delay, scaling laws, multihop networks, mobility models.

*E-mail addresses: vassilis@netmode.ntua.gr
†E-mail addresses: papavass@mail.ntua.gr

1. Introduction

Wireless networks have become a significant part of modern networking systems, both as access interfaces and core infrastructures. Various wireless network types have been devised for different purposes and applications, leading into two main categories. Centralized networks with fixed infrastructure, like cellular and wireless LANs on one hand, and unstructured environments like ad hoc and sensor networks on the other.

Cellular networks differ fundamentally from their ad hoc counterparts, especially with respect to the routing of data in the network. Packets undergo only one-hop transmissions in order to reach the base station, whereas in unstructured networks, packets need to traverse multiple hops of wireless links to be delivered to their final destination. For this reason the latter category of networks is oftentimes referred to as multihop networks.

One of the main functionalities of ad hoc and sensor networks, namely using many collaborating low-cost nodes to perform a greater task, often calls for the basic network units to move around the deployment area. This is mostly desired in the cases where the nodes are attached to vehicles, carried by pedestrians or military troops, or they constitute robotic units. At the same time, mobile nodes need to exchange data, potentially at high data rates, a prominent problem even for static wireless nodes. Mobility and information transmission need to be performed efficiently in terms of energy consumption, often under stringent environmental requirements. The quantities of interest for evaluating such communication capabilities are the network traffic-carrying capacity, node throughput, packet loss, packet delivery delay and energy consumption.

The purpose of this chapter is to study and discuss the impact of mobility on wireless multihop networks with respect to the aforementioned performance metrics. More specifically, after a brief presentation of various commonly used mobility models, the chapter focuses on the inherent tradeoff between capacity and packet delay parameters. This tradeoff emerges as a consequence of the mobility capability in wireless multihop networks, in which case it can be used to increase the traffic-carrying capacity of the network by exploiting proper relay transmission policies. On the other hand, such mobility-exploiting strategies may increase the delay of delivered packets with negative consequences for some networking applications, especially real-time and QoS-oriented ones. This three-way relation is the main subject of this chapter and it is explored from various perspectives.

The rest of this chapter is organized as follows. In section 2., various frequently used mobility models are presented and their basic characteristics are provided. In section 3., the aggregate capacity of various network arrangements is presented and asymptotic characterizations of their scaling behavior are provided. In section 4. the tradeoff between capacity and delay that emerges as a consequence of mobility in wireless multihop networks is described and analyzed extensively, while in section 5. a qualitative and quantitative comparison of the various approaches is presented. Finally, in section 6., some conclusions are drawn and further directions for future work are provided.

2. Mobility in Wireless Multihop Networks

A significant number of mobility models have been proposed in the context of wireless networks. In general, these models can be classified in two main categories, namely random and correlated ones. Several correlations may be considered for the latter, most notably temporal, and spatial. Figure 1 summarizes the mobility models described in this chapter and their typical application environment.

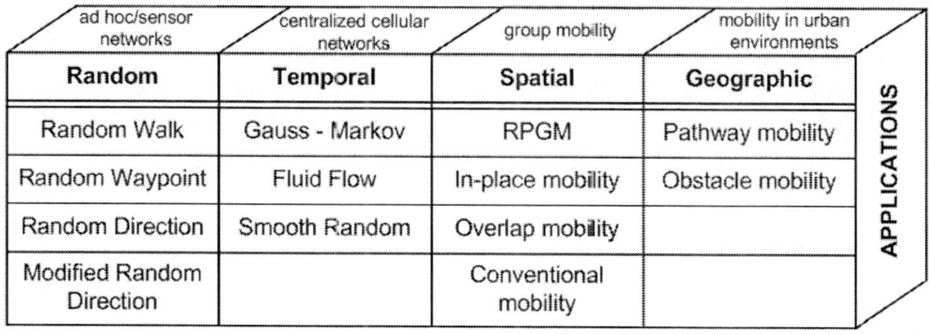

Figure 1. Classification of mobility models and typical applications.

2.1. Random Mobility Models

The simplest model is the Random Walk (RW), also referred to as *Brownian Motion*. In RW, the speed and direction are picked uniformly from the ranges $[v_{\min}, v_{\max}]$ and $(0, 2\pi]$ respectively. Reflection or wrapping at the edges of the deployment region may be employed. For both flavors, the node spatial distribution in the steady-state is uniform [1]. The RW model may be considered as a memoryless stochastic process, and consequently it is not very realistic, as the entities employing it exhibit sharp angle changes, instantaneous velocity variations and no pausing intervals.

An immediate extension is the Random Waypoint Model (RWP), which is the most frequently used in wireless networks. A node employing RWP chooses uniformly at random a destination point (waypoint). It then moves to the destination with a speed, which is a uniformly distributed random variable in the interval $[v_{\min}, v_{\max}]$. When it reaches the destination, it pauses for a period of t_p in the interval $[t_{p\_\min}, t_{p\_\max}]$, where $t_{p\_\min} > 0$. The procedure is repeated after each pause. The random selection of the destination can be done, by choosing a destination direction, which is uniformly distributed in the interval $(0, 2\pi]$ with respect to a predefined physical reference direction [2]. The resulting spatial node distribution consists of three components, the static, the pause and the mobility component, from the analytical expressions of which some very important results emerge. The distribution is independent of the node speed and the initial spatial distribution. More important is the fact that the initial distribution is not preserved. Therefore, one may initially deploy the nodes in a scenario according to the steady-state distribution, in order to acquire from the beginning more accurate results.

In Random Direction (RD), a node moves in a constant direction, at a constant speed until it reaches the boundary, where it stops for a time interval of duration t_p. If a node is allowed to stop and choose a new direction before it reaches the boundary the model is called Modified RD. The speed and direction in different intervals are not related to their past values, thus allowing nodes to have sharp turns or sudden stops [3,4]. A simple variation to the above model is to consider that the terminal has a constant velocity and its direction is regenerated every time interval. Another variation is to force the terminal move along a piecewise path. The segment lengths of the path are exponentially distributed and the direction is randomly chosen. Speed is randomly distributed between $[v-\Delta v, v+\Delta v]$, Δv being the maximum unit vehicular change.

Random mobility models are not very realistic, however they can effectively capture independent node movement. Thus, they are appropriate to model mobility in wireless ad hoc and sensor networks, where central infrastructure is absent and nodes operate in a peer-to-peer and opportunistic fashion.

2.2. Temporal Dependence Models

Such models accurately describe mobility in centralized wireless networks, like cellular and WLANs. The Gauss-Markov model is the most popular one and it constitutes a predictive, distance-based scheme. The velocity of a node is modeled as a Gauss-Markov stochastic process (correlated over time) [5]. A node hitting the boundary of the deployment region is reflected back by changing the direction of the movement by 180 degrees. By adjusting the memory level of the process m (scalar parameter), several mobility models may be yielded with various degrees of temporal dependencies. When $m = 0$, the model becomes the basic RW. As m increases, the current velocity is more likely to be influenced by the previous one. When $m = 1$, the model becomes the fluid flow model described in the sequel.

Fluid flow models describe motions that resemble the flows of fluids [6–8]. In [6], three fluid flow models are studied. In the first, *model A*, the speed and direction are independent random variables and they are unchanged. The direction is uniformly distributed in $(0, 2\pi]$. In *model B*, the speed and direction are again independent random variables. They are regenerated randomly and independently when the terminal crosses a boundary of a cell. The new direction is in $\left(-\frac{\pi}{2}, \frac{\pi}{2}\right)$ relative to the previous direction of the terminal. Finally, in *model C*, the speed and direction are again independent random variables, however, they are regenerated after a random time interval or a random traveling distance. The direction is randomly and uniformly selected in $(0, 2\pi]$, the random time interval has mean T and the random distance has mean D. Under mobility equilibrium conditions, model C, like model A, does not favor any specific location. Thus, both models distribute the terminals uniformly across a cell. For model B, even though the above is not true, if one employs the distributions of model A for the new speed and direction for an incoming terminal into a cell, then from the cell's perspective, there is no difference between the two models. Thus, the terminals can still be uniformly distributed everywhere.

In reality node speeds tend to obtain a certain range of preferred values, rather than purely random ones. In the Smooth Random mobility model [9] the probability distribution of a node's velocity assigns a high probability to the set of preferred speeds, assuming a uniform distribution on the remaining part of the interval $[0, v_{\max}]$. The frequency of the speed change is considered to be a Poisson process. The speed is incrementaly/decrementaly changed at speed change events, with uniformly distributed acceleration/deceleration. For each time slot the new speed is calculated as $v(t) = v(t-\Delta t) + \gamma(t)\Delta(t)$, where $\gamma(t)$ denotes acceleration/deceleration. The latter also defines the strength of the temporal correlation, as the greater $\gamma(t)$ is, the stronger the temporal correlation. Movement direction is purely random and it is uniformly distributed in the interval $(0, 2\pi]$. The frequency of direction change is assumed to have an exponential distribution. Even though the direction change can be large, the actual change takes place smoothly by dividing the selected direction change into several incremental smaller direction changes $\Delta\phi(t)$ per time slot. The $\Delta\phi(t)$ also determines the temporal correlation degree of the model. By adjusting these two parameters, different mobility models may be obtained with variable degrees of temporal dependency.

2.3. Spatial Dependence Models

In such models, mobility is defined according to spatial parameters. The most popular model, the Reference Point Group Mobility (RPGM) is a group mobility model, where each group has a logical center (similar to the center of mass), which could also correspond to a physical leader node and whose motion defines the group's motion behavior [3]. A path for the center determines the trajectory of the group. Assuming the nodes are uniformly distributed around the center of a group, assigning a reference point to each node, such that the point follows the group motion, a node can move independently in addition to the group motion (i.e. swarming [10]). The path that a group follows is given in a form of checkpoint sequence with possibly additional time constraints. As time evolves, a specific group moves from one checkpoint to another. Once it reaches the new point, it computes the new motion vector. This allows decoupling the motion pattern from the model itself.

By proper selection of check point path and initial group location various scenarios may be modeled based on RPGM. In the *In-place* mobility model, the entire area is divided into adjacent regions with different groups in each region. The mobility patterns of each group can be rather different. In the *Overlap* mobility model, the various groups work in the same area and follow different patterns, following an overlapping manner. Finally, in the *Convention* mobility model, the situation resembles that of exhibitors and attendees in a convention hall. Groups of nodes roam from region to region. Groups might stop or pass through each of the visited region.

2.4. Geographic Restriction Models

Compared to independent mobility models, mobility in urban environments is significantly different. Specially adapted models have been devised to compensate for such discrepancy and provide a more accurate urban mobility description. In the pathway mobility model [11], nodes move according to pre-defined pathways of a map in the network terrain. The map is a network graph where vertices represent generic obstacles and the edges the freeways or streets of the arrangement. A random graph may be employed for modeling the map of a city. Initially, nodes are randomly placed on the vertices of the graph and destinations are randomly chosen, while nodes move towards the selected destinations via the shortest paths of the map graph. Nodes pause at each destination for t_p and the process is repeated for a new destination point.

In realistic scenarios, obstacles play an important role in the trajectory of nodes. Obstacles may be taken into account as rectangular boxes randomly placed on the considered field. In the obstacle mobility model [12], node trajectories are required to avoid obstacles and also the transmitted signal may be influenced (either attenuated or fully absorbed) by the encountered obstacles. The terrain can be modeled as a graph in a similar way as in pathway mobility models. Buildings and obstacles are taken into account in order to construct the allowed pathways, and nodes are allowed to move only along these pathways. Consequently, nodes also travel in a pseudo-random fashion as in pathway mobility, following the shortest path between their current point and the selected destination. Several mobility scenarios maybe customarily produced in this way, such as conference room (most nodes static, a small percentage mobile), event coverage (large groups of highly mobile users) and disaster relief scenarios (some users slow, the rest highly mobile).

3. Capacity in Mobile Multihop Networks

Continuous and random node movement can influence the operation and performance of localized network protocols that usually rely on the collection of local neighborhood information. Asynchronous sampling of this information, affects protocol operation in various networking layers and can lead to two mostly undesirable side effects, namely inconsistent network view and outdated local information [13].

In typical localized mechanisms, the collection of neighbor information is performed through periodic, asynchronous "HELLO" messages. Such information is denoted as *local view*. Assuming a connected initial network topology, a node uses the local view information to select a subset of nodes from the one-hop neighbor set. This subset is called *logical neighbors* and the union of all these subsets *logical topology*. When mobiles are allowed to move, some logical neighbors might not be reachable after a while. Reachable nodes are called *effective neighbors*. The union of the effective neighbor sets forms the *effective topology*. Both logical and effective topologies might become disconnected due to node mobility [13]. A disconnected logical topology is the result of inconsistent local views. Such inconsistencies come from the fact that nodes use the same information to build their local view, which results in inaccurate global view (union of local views). Outdated information can lead to disconnected effective topology, as well. Assuming all nodes built their local views consistently through a synchronization mechanism that allows them to use the same network information, movement of nodes after the global view is created and before the next update instant, creates similar situations as in the case of inconsistent local views.

The loss of links in the effective topology leads to very undesirable consequences for the network performance. The most salient is packet dropping. Network throughput is severely impacted, as the packets destined to a node have to be resent multiple times until the link failure is discovered. Similarly, packet delay increases, due to the buffering of undelivered packets. However, even though mobility is responsible for such undesired behavior in wireless multihop neighbors, it could also have some unexpectedly pleasant effect on performance of such networks.

3.1. Capacity of Static Wireless Multihop Networks

Capacity is one of the fundamental performance metrics, indicating the total information transporting capability of a network. In the following, we start by mentioning some fundamental capacity results obtained for static wireless networks. These results are usually employed as benchmarks for comparison purposes among the various protocols proposed for wireless mobile networks.

Achieved results for static ad hoc networks are of asymptotic nature, namely limiting expressions as the number of network nodes grows to infinity [14]. Assuming a unit-disk area deployment region, nodes are spread over the terrain independently and uniformly at random. Random node locations $\{X_i\}$ are fixed over time. Nodes act simultaneously as sources, destinations and relays for multihop traffic. The destinations are independently and randomly chosen in the network, while all nodes are assumed to have the same transmission power P, while the channel gain from node i to node j has the form $g_{ij}(t) = |X_i(t) - X_j(t)|^{-\alpha}$, $\alpha > 2$ (α being the path loss exponent). The transmission from i to j will be unsuccessful when there is an interferer k within distance $|X_k - X_j| \leq \left(\frac{\xi}{G}\right)^{1/\alpha} |X_i - X_j|$, where ξ is the SIR requirement of a node for successful communication and G the processing gain of the system.

For the system model described above, it is possible to obtain upper and lower bounds on the asymptotically feasible throughput $T(n)$ for a network with n nodes [14]. Specifically, it has been shown that there exist constants c and c', such that

$$\begin{aligned} \lim_{n\to\infty} \Pr\left\{ T(n) = \frac{cR}{\sqrt{n\log n}} \text{ is feasible} \right\} &= 1, \\ \lim_{n\to\infty} \Pr\left\{ T(n) = \frac{c'R}{\sqrt{n\log n}} \text{ is feasible} \right\} &= 0 \end{aligned} \quad (1)$$

Successful transmission between i and j requires absence of interferers in a disk of radius proportional to the transmission distance $|X_i - X_j|$. This means a successful transmission over a distance d incurs a cost proportional to d^α by excluding other transmissions in the neighborhood of source i. Thus, the capacity is constrained by the mutual interference of concurrent transmissions. In light of this result, one can maximize the transport capacity by scheduling large numbers of short transmissions, instead of small numbers of longer ones. The best approach then, is to restrict transmissions in distances of the order of $1/\sqrt{n}$. In that case, the transport capacity will be at most $\sqrt{n}$ (bit $\times$ m/s). In this scheme, there are n sessions with an expected distance of $\Theta(1)$[1], yielding a throughput of at most $O(1/\sqrt{n})$ per session. As n increases asymptotically, the throughput per source-destination (S-D) session decreases as $1/\sqrt{n}$ and this is the best achievable result under an optimal scheduling, routing and relaying policy [14]. If the transmission ranges are constrained to the order of $1/\sqrt{n}$, the number of intermediate hops from source to the destination is $\sqrt{n}$.

[1]For the rest of the chapter, given two nonnegative functions $f(n), g(n)$, the following notation is employed:

- $f(n) = O(g(n))$, if there exist constants $c, m > 0$ such that $f(n) \leq cg(n)$, for every $n > m$
- $f(n) = \Omega(g(n))$, if $g(n) = O(f(n))$
- $f(n) = \Theta(g(n))$, if $f(n) = O(g(n))$ and $g(n) = O(f(n))$
- $f(n) = o(g(n))$, if $\lim_{n\to\infty} \frac{f(n)}{g(n)} = 0$
- $f(n) = \omega(g(n))$, if $g(n) = o(f(n))$

3.2. Mobility Increases the Capacity of Ad Hoc Networks

The interference levels influence significantly the capacity of a static ad hoc wireless network [14]. Essentially, such a destructive factor influences the channel quality. Together with multipath fading, obstacle shadowing and path loss distance attenuation, these factors are responsible for the time variations of signal strength of a wireless link. The time variation of the channel in turn, constrains the available transmission intervals, thus directly affecting the throughput of the network.

An effective framework has been developed for analyzing mobility and capacity in wireless ad hoc networks and an intelligent approach for utilizing the first in favor of the second has been developed in parallel [15, 16]. Multiuser diversity is proposed for achieving that goal. In a system with many users, whose channels vary independently, it is likely at any time to have a user with very high channel quality. This user should be allocated the common channel resource, exploiting channel gain at maximum. Utilizing the result acquired for static ad hoc networks, one can suggest that in order to create diversity a source has to wait until it is close to the destination, at a distance of the order of $1/\sqrt{n}$ to transmit. Unfortunately, the fraction of time two nodes are nearest neighbors is of the order of $1/n$, thus limiting the available capacity as in the static case.

A different approach is taken in order to improve the traffic-carrying capacity of a mobile network [15, 16]. Considering a deployment region having the shape of a unit-area disk, and denoting node density by n (equal to the number of network nodes), the location of a mobile at time t is given by $\{X_i(t)\}$ (stationary and ergodic process). Node motions are considered random, independent and identically distributed. The transmit power of each node at time t is denoted by $P_i(t)$. With channel gain g_{ij} and an SIR requirement for successful communication of ξ, a transmission rate $R(\xi)$ from i to j at time t can be achieved if

$$\frac{P_i(t)g_{ij}(t)}{N_0 + \frac{1}{G}\sum\limits_{k \neq i} P_k(t)g_{kj}(t)} > \xi \tag{2}$$

where N_0 is the background noise power and G the processing gain of the system. Each node is assumed to have an infinite buffer to store relayed packets and the applications do not impose any delay constraints. A centralized scheduling policy is considered, which decides on the nodes that are going to transmit in any slot (which packets from the queue of node i to transmit and what power to use for each packet transmission). The long-term throughput of the system is denoted by $T(n)$.

In the mobile network, any two nodes are expected to be close to each other from time to time. Consequently, capacity may be improved by preventing relaying of packets and allowing the sources to transmit directly to their destinations, when distance permits that. This policy achieves a $\Theta(1)$ throughput per S-D pair [16]. Indeed, assuming that the policy π is allowed to schedule direct transmission between the source and destination with no intermediate relaying, if c is any constant satisfying

$$c > \left[2^{\alpha}\left(1+\frac{2}{\alpha}\right)\pi^{-\alpha/2}\frac{\xi+G}{\xi}\right]^{\frac{1}{1+\frac{\alpha}{2}}} \tag{3}$$

then

$$\Pr\left\{T(n) = cn^{-\frac{1}{1+\frac{\alpha}{2}}}\right\} = 0 \tag{4}$$

for sufficiently large n. Equation (4) implies that without relaying the achievable throughput per S-D pair goes to zero at least as fast as $n^{-\frac{1}{1+\frac{\alpha}{2}}}$.

The above result assures that for the same scheduling policy and a fixed time instant t, if $S(t)$ is the set of source nodes that are scheduled successful transmissions to their respective destinations, then

$$\sum_{i \in S(t)} |X_i(t) - X_{d(i)}(t)|^{\alpha} \leq B \tag{5}$$

for

$$B \doteq 2^{\alpha}\pi^{-\alpha/2}\frac{\xi + G}{\xi} \tag{6}$$

Equation (5) expresses the fact that the number of simultaneous long-range communication is limited by interference. This puts a limitation on the performance of any strategy which uses direct transmission, since the S-D distance in the model is of the order of $\Theta(1)$. Together (6) and (5) suggest that if transmissions over long distances are allowed, there would be several S-D pairs that are within range. On the other hand, if communication is constrained in short distances, there is only a very small fraction of S-D pairs that are close enough to transmit packets. Consequently, the throughput of the network is interference (first case) and distance limited (second case).

Moving one step forward would be to allow local communication, to comply with the interference requirements, while permitting enough S-D pairs with packets to transmit. The way to achieve the first is straightforward by reducing node transmission power. The second implies the use of relaying. Intuition leads to a greedy forwarding scheme for relaying packets, in which traffic is spread to a large number of intermediate nodes, so that in the steady-state, the packets of all sources are distributed across many nodes in the network (if possible). That way it is likely that every node in the network will have packets stored destined to all other nodes except itself, so that a scheduled S-D pair always has a packet to send.

Since the node location process is stationary and ergodic and the sample random functions $X_i(t)$ are independent, it suffices to relay each packet only once. That is, the packet stream is split across different routes in the network, instead of copying each packet and distributing the copies across the routes. The reason for this is that the probability for an arbitrary node to be scheduled to receive a packet from a source node S is equal for all nodes and independent of S. Each packet making two hops, is transmitted no more than twice and the available total throughput is $\Theta(n)$.

A two-phase policy π for increasing capacity (henceforth this policy will be referred to as Basic Relay Model) has been proposed in [15, 16]. In a time slot t, a sender density parameter $\theta \in (0,1)$ is defined. The node set V is split in a set S of senders with cardinality $n_S = \theta n$ and the remaining set W with cardinality $n_W = n - n_S$ of potential receivers. In the second phase, all senders transmit packets to their nearest neighbors among all nodes in W using unit transmit power. Namely, in each time slot, each selected sender will transmit packets to the closest receiver node. By construction, there are n_S sender-receiver pairs.

However, not all of them are feasible, due to interference. Denote with n_t the number of feasible S-D pairs among the n_S potential pairs.

Interesting and useful results can be obtained within the aforementioned framework. The first is that the expected number of feasible S-D pairs is $\Theta(n)$. Namely,

$$\lim_{n\to\infty} \frac{E[n_t]}{n} = \phi > 0 \tag{7}$$

where ϕ is a constant. Moreover, for two arbitrary nodes i and j, the probability that (i, j) is a scheduled S-D pair is $\Theta(1/n)$. Every packet goes through at most two hops, since at the first phase a source transmits either to a relay or a destination and at the second phase a relay transmits to the destination.

Relation (7) is very important. It is the base to prove that the two-phased algorithm given above achieves a throughput per S-D pair of $\Theta(1)$. This means there exists a constant $c > 0$ such that

$$\lim_{n\to\infty} Pr\{T(n) = cR \text{ is feasible}\} = 1 \tag{8}$$

with maximum possible throughput equal to $c = \frac{\phi}{2}$.

By expression (7) it can be deduced that the long-term throughput between any two nodes is equal to the probability that these two nodes are selected by policy π and it is of the order of $\Theta(1/n)$. For a given S-D pair, there is one direct route and $(n-2)$ two-hop routes, through a relay node. According to the above, the throughput of the direct route is $\Theta(1/n)$. If the two-hop routes through a relay are considered a single server queue, both the arrival and service rate of the queue is $\Theta(1/n)$, with reference to (7). Summation of all the throughputs of the $(n-1)$ routes yields a total throughput per S-D pair of $\Theta(1)$. This essentially leads to (8). The result in (8) is possible, due to the fact that the received power at the nearest neighbor is of the same order as the total interference from $\Theta(n)$ interferers. This allows for $\Theta(n)$ concurrent nearest neighbor transmissions.

A final refinement of the policy is to make the scheduling algorithm distributed. At any time instant, each node can decide whether it wants to be a sender or a receiver. It seeks out potential receivers nearest to it and attempts to send them data. By numbering time slots and phases sequentially, in an even phase, senders forward packets from sources to relays and in odd slots packets are forwarded from relays to destinations. An important note is that the procedure of senders selecting the closest receiver in the scheduling policy can be seen inversely as well, with a receiver selecting the closest sender from which to receive. This approach is called receiver-centric, compared to the sender-centric presented above. However, the two approaches are not symmetrical. In the receiver-centric approach, a receiver is likely to receive from multiple senders. Then all packets but one will have to be discarded based on the specific strategy.

Conclusively, the key idea is to exploit mobility of nodes in ad hoc wireless networks. Starting from the basic result for static networks, namely that the capacity goes asymptotically to zero as the node density increases, multiuser diversity is used by relaying packets only once, in distance-close neighbors. Owe to mobility, it is very likely that the intermediate nodes will eventually be close to the final destination and deliver their packets. This way a per S-D throughput is reached, which is constant as the node density increases asymptotically. The benefit of mobility is striking, since by employing the previous policy total throughput scales linearly with the number of nodes, as opposed to sublinearly for the case of static networks.

3.3. Effects of High Mobility

The previous two results have assumed perfect channel knowledge. However, high degrees of mobility complicate system behavior in two ways, namely channel uncertainty and network homogeneity [17]. In this case, wireless channels between mobile nodes vary rapidly and reliable channel estimation is impossible. Mobility complicates the distinction between nearest neighbors and distant nodes as the topology of the network is constantly changing. Above a specific degree of mobility nodes appear statistically equivalent. Network homogeneity refers to the statistical equivalence of the channels between different pairs of nodes resulting from high node mobility and the inability of nodes to track the rapidly changing network topology. Network homogeneity can be both an advantage (as in the case of connectivity) or a disadvantage as explained before. Especially in the case of MIMO networks, mobility leads to the additive results of loss of degrees of freedom. Multiple users and multiple antennas offer additional degrees of freedom in the spatial domain. However, with insufficient channel information, the degrees of freedom and respective capacity benefits are lost.

It has been shown in [17] that for the multiple-antenna ad hoc network increasing the number of transmit antennas beyond the channel coherence time does not increase the capacity, due to the aforementioned implications of mobility. In terms of homogeneity, when considering a rapidly varying channel, and given the inability of the transmitters and receivers to track the topology or the fading channel state information (CSI), the channels between any pair of users can be modeled as isotropic fading, memoryless, independent and identically distributed, stationary, ergodic stochastic processes with realizations unknown at both ends. The exact capacity region of such ad hoc network is determined given any partition of nodes into source, destination, and relay types. As before, all degrees of freedom are lost due to insufficient channel knowledge, and in these highly mobile networks, there are no capacity benefits of any cooperation scheme such as relaying between nodes. In essence, it is shown that with too much mobility, the throughput of the ad hoc network does not depend on the number of users except through the total transmit power. In fact, total network throughput grows no more than double logarithmically with the number of nodes [17].

When the coherence time is long enough so that the receiver is able to track the channel variations perfectly, even though the capacity region of ad hoc networks in these cases is not known, it is easily seen that the total throughput cannot increase more than logarithmically with the number of nodes. Thus, in the limit of very high mobility, the total throughput of the ad hoc network grows lower than that of a fixed wireless network even if the receiver is able to track the channel.

3.4. Capacity of Restricted Mobility Networks

In the mobility pattern considered in the Basic Relay model, node trajectories are considered independent, stationary and ergodic random processes with uniform stationary distribution on the unit disk. Such model is however unrealistic in several actual settings. Nevertheless, throughput can be kept constant even in the event that mobility is restricted in a single dimension [18]. More specifically, even if node mobility is restricted in uniformly and randomly selected great circles on a sphere it is shown that there exists a relaying policy achieving a constant throughput $T(n) = \Theta(1)$ per S-D pair.

A different type of restricted one-dimensional mobility may be considered in which, the unit-area deployment region of the network model in the Basic Relay scheme is partitioned into square cells and nodes are restricted to move within one randomly chosen cell [19]. Two cases are considered in which the cell area either scales as $\log n/n$ or remains constant, obtaining results close to that of fixed-location networks and that of the Basic Relay model, respectively. In this case, a scheme is proposed, where only transmissions to closest neighbors are allowed. The scheme relies on multihop routing for information transferring between S-D pairs. It is shown that in the case of constant cell size $\frac{1}{\ell}$, throughput scales as $T(n) = \frac{1}{\sqrt{\ell}}\Theta(1)$, while for the case of cell sizes varying with the number of nodes n, the associated throughput is $T(n) = \Theta\left(\sqrt{\frac{\log n}{n}}\right)$.

3.5. Capacity of General Heterogeneous-Mobility Nodes

Previously described asymptotic results strongly rely on the assumption that nodes are identical, in the sense that they move according to some ergodic process and it is equally likely to visit any portion of the network area. However, this is more a convenient approximation to the actual behavior of mobiles, rather than a realistic assumption. In reality nodes follow different mobility patterns from each other and a general framework to characterize the capacity of networks with such arbitrary mobility patterns would be useful and more practical. Such a description is developed in [20], considering both the case of finite number of nodes, while asymptotic results in the event that the number of nodes grows to infinity are also provided.

The general case of heterogeneous mobile nodes excludes hybrid networks since they constitute a different network type, consisting of nodes with different capabilities. On the contrary in networks with heterogeneous mobiles, the nodes have the same networking capabilities but follow different mobility patterns. The general case of heterogenous mobiles can be initially formulated as a joint scheduling and routing problem, defining an abstract framework with respect to the characterization of network traffic by means of the traffic matrix, within which the performance analysis of mobile ad hoc networks can be carried out. According to this formulation, a precise characterization of the capacity region of a network with finite number of nodes may be obtained and extended to the description of several structural properties of the system. The proposed methodology defines explicitly the acceptable traffic combinations (i.e. sustainable traffic matrix) and defines the corresponding capacity region of the network. Utilizing the notion of contact graphs[2], the routing problem can be formulated as a multi-commodity flow problem on a contact graph [20]. A rather interesting result is that the class of simple policies achieves maximum throughput (i.e. no gains can be obtained by adopting complex policies that schedule transmissions by

[2]A graph whose vertices correspond to network nodes and its edges to communications links according to an employed scheduling policy S.

considering dynamical variables such as instantaneous queue lengths, age of stored information at the nodes, etc.), while the capacity region depends on the mobility model process only through the joint-stationary node distribution. The capacity region of an ad hoc network with mobile nodes is convex. Similarly, the distribution of edge capacity (i.e. that of edge links) is characterized by heavy tails [20].

Furthermore, the aforementioned analysis can be easily extended to the asymptotic regime. For a number of nodes approaching infinity it is proven that a throughput of $T(n)$ is achievable only by those scheduling policies that select with high probability (w.h.p.) transmitter and receiver pairs with distance of the order of $O\left(\frac{1}{\sqrt{f(n)}}\right)$. Furthermore, for independent node movements and finite asymptotic node density (different than zero) the asymptotic capacity is achievable by the class of scheduling policies which force the transmission range to be w.h.p. of the order of $O\left(\frac{\log n}{\sqrt{n}}\right)$. Overall, in an anisotropic mobility pattern application environment, capacity may vary in a continuous fashion from $\Theta(\sqrt{n})$ (fixed-location networks case) to $\Theta(n)$ (Basic Relay model case).

On the contrary, for the case of actual hybrid networks consisting of wireless ad hoc nodes and randomly placed base stations (i.e. mesh networks with increased complexity), where the base stations are equipped with high-bandwidth capabilities and do not act as sources or destinations, the capacity strongly depends on the number of these base stations [21]. For a hybrid network of n nodes and m base stations, analytical results show that if m grows asymptotically slower than $\sqrt{n}$, the benefit of adding base stations on capacity is insignificant. However, if m grows faster than $\sqrt{n}$, the throughput capacity increases linearly with the number of base stations, providing an effective improvement over a pure ad hoc network. Therefore, achieving meaningful capacity gains, requires a considerable wired infrastructure and respective financial cost.

4. Mobility-Induced Capacity-Delay Tradeoff in Wireless Ad Hoc Networks

The Basic Relay Model (section 3.2.) requires packets to be buffered at intermediate relay nodes until the channel becomes strong (relay is close to the destination) and a packet is ultimately delivered. Various techniques in the literature adopt this approach and define more complex relay schemes. It becomes apparent that asynchronous strategies of this type incur additional delay, due to packet buffering and emerging bottlenecks at relays as traffic loads increase. In the following, we present various approaches analyzing the delay effects of mobility and their implications on capacity.

In a fixed network, the delay equals the sum of the times spent at all the relays. In a mobile network, delay is given by definition by the same sum, however, it will now depend on the velocity $v(n)$ that each relay has, since packet routes may now alter due to mobility. Considering the same network model as for the case of static networks and the Basic Relay model, it is possible to study the delay in a rigorous framework and obtain analytical results similar to those acquired for the capacity. Indeed, for a static network on a grid of square cells each of area $w(n) \geq 2\log n/n$ and achievable throughput $T(n) = O(1/\sqrt{n\log n})$, delay scales as $D(n) = \Theta(nT(n)) = \Theta(\sqrt{n/\log n})$ [22]. Similarly, for a mobile network under the framework of the Basic Relay model (network area is partitioned into n cells each of size $1/n$), throughput scales as $\Theta(1)$ and assuming relay node queues are of the GI/GI/1-FCFS type, the delay scales as $D(n) = O(\sqrt{n}/v(n))$. Two scheduling schemes may be employed for achieving the optimal throughput and delay values in ad hoc networks. In the first, variable size cells are assumed and a TDMA scheme is defined for scheduling which cell will be active and its nodes will transmit within it or in neighboring cells. A source transmits to its destination along adjacent cells lying on the straight line connecting the source to the destination. Active cells transmit a single packet for each S-D line passing through it. In the second scheme, cells are of constant size and in active cells transmissions take place only within the cell. The active slot of a cell is further divided in two sub-slots, in the first of which a randomly selected node transmits a packet to a destination (if present), otherwise to a randomly selected node within the cell. In the second sub-slot a randomly selected node transmits directly to another randomly selected node within the cell.

Even in the case of restricted mobility, where nodes move across randomly and uniformly chosen great circles on a sphere [18], throughput and delay performance seems to be unaffected by mobility restrictions. Indeed, for such a system model and a scheme similar to the second presented above, a throughput of $T(n) = \Theta(1)$ and a delay of $D(n) = \Theta(\sqrt{n}/v(n))$ may be attained [23].

4.1. Capacity-Delay Tradeoff-oriented Schemes

The capacity-delay tradeoff that emerged from the previous analysis is very critical for the performance of wireless multihop networks and design of new protocols. Several approaches have been presented that address this tradeoff and numerous solutions for balancing it have been proposed. Depending on the details of each specific system model, asymptotic results may differ, however, they are all compatible with the basic elements expressed by the capacity-delay tradeoff.

The analysis and results of the Basic Relay model revealed that delay can be decreased by increasing the size of the neighborhood of each node. However, a larger neighborhood would result in lower throughput due to increased interference, thus the emergence of the aforementioned tradeoff. Intuitively, to obtain higher network throughput, one needs to use mobility, and to obtain lower delay, multiple hops need to be used cleverly. In [22] this tradeoff is addressed by a multihop relay policy, in which the region is partitioned in cells of area $w(n)$. An additional grid formed by square sub-cells of size $b(n) = \Theta(\log n/n)$ is created, so that each square cell of area $w(n)$ contains $w(n)/b(n)$ sub-cells each of area $b(n)$. The time slot of each cell is divided into $\Theta(nw(n))$ packet time slots, forcing packet sizes to scale as $\Theta(1/nw(n))$. Packets are delivered to their final destination via multihop relaying. It is shown that if the velocity scale rule $v(n) = o(\sqrt{\log n/n})$ is satisfied, the above scheme achieves $T(n) = \Theta\left(\frac{1}{\sqrt{nw(n)\log n}}\right)$ and $D(n) = O\left(\frac{1}{v(n)\sqrt{w(n)}}\right)$, where $w(n) = O(1)$ and $w(n) = \Omega(\log n/n)$, where $\sqrt{w(n)}$ denotes the average distance traveled by a mobile in one hop. The scheme achieves the optimal throughput-delay tradeoff by varying the number of relay hops and the transmission range.

In the following we present some additional approaches that address the capacity-delay tradeoff and they also consider slightly different system models with respect to mobility and analysis.

4.2. Packet Redundancy Approach

An alternative approach suggests adding redundancy in the packet transmissions by sending copies of packets along multiple paths in the network, reducing the end-to-end delay at the expense of reduced total network throughput [24]. Redundancy alone cannot increase network capacity, but it can significantly improve delay performance, yielding delay reductions by several orders of magnitude when data rates are sufficiently less than capacity.

The deployment region (assumed square-shaped) is divided into C square overlapping cells of equal size. A total of n mobile nodes drift around the cells and the time is considered slotted. Two nodes can transmit to each other when they are in the same cell during a time slot. Then, a cell can only support one transmission per time slot. Mobiles potentially change a cell after a time slot. The chosen mobility model, allows a node to choose its new cell in an independent and identically distributed manner. Nevertheless, it is shown that the obtained capacity results are valid even for simple Markovian random walks where users move to adjacent cells. The independent and identically distributed mobility model essentially sets a bound on the case of infinite mobility performance. In order to avoid inter-cell interference, each cell is assigned a different channel frequency and only four different frequencies suffice for a full legitimate assignment (four-color theorem from graph theory).

The arrival rate of packets at node i is denoted by λ_i, and the arrival process is assumed to be Bernoulli. Each source has a specific destination. In [24] a cyclic transmission scheme was assumed. The arrival rates of the nodes have to be such that the queue sizes do not grow infinitely. Then the capacity of the network is the maximum rate λ among all λ_i that the network can support stably. Relaxing the problem and assuming $\lambda_i = \lambda$, the capacity of the network, from a network layer perspective (where the optimization is over all possible routing and scheduling protocols), is $\mu = \frac{p+q}{2d}$ with $p = 1 - \left(1 - \frac{1}{C}\right)^n - \frac{n}{C}\left(1 - \frac{1}{C}\right)^{n-1}$ and $q = 1 - \left(1 - \frac{1}{C^2}\right)^{n/2}$, the first one giving the probability of finding at least two users in a particular cell, while the latter gives the probability of finding a source-destination pair within a cell. The network can support simultaneous communication rates $\lambda \leq \mu$. The asymptotic capacity as n goes to infinity is $\frac{1-e^{-d}-de^{-d}}{2d}$, where $d = n/C$. This value goes to zero as d goes to either zero or infinity. This is rather reasonable as when d tends to zero the probability of finding two users in a cell diminishes to zero as well. When d approaches infinity, there will be so many users in a cell, that most of them will be idle when an S-D pair is selected.

When heterogeneous rates λ_{ij} are used (where λ_{ij} represents the rate user i receives exogenous data intended for user j) and K is the maximum number of destination users to which a source transmits, then the symmetric capacity region is the region of all stabilizable data rates, such that no user is receiving or transmitting at a higher total rate than any other node. Then, the symmetric region has the form $\sum_j \lambda_{ij} \leq \frac{1-e^{-d}-de^{-d}}{2d} + O(K/n)$ for every i and $\sum_i \lambda_{ij} \leq \frac{1-e^{-d}-de^{-d}}{2d} + O(K/n)$ for every j.

A relay algorithm, similar to the two-hop relay scheme of the Basic Relay model is proposed for achieving capacity. Again, the use of relay nodes is very important as it allows throughput to be limited only by the rate at which a source encounters other users, rather than by the rate, it encounters the destination node. At any time slot, for every cell with more than two users, if there exists a feasible S-D pair within a cell, a random pair is chosen among those pairs and the source transmits a new packet to the destination, otherwise it remains idle. If there is no S-D pair, a random node is chosen as sender, and another one as receiver. With equal probability, either a relay packet is sent from the chosen source to the corresponding destination, if any, or a new relay packet is sent from the selected source to the destination. If none of the above is feasible, the source remains idle.

For the system described thus far, with the mobility model of users changing cells independently and identically distributed in a uniform fashion at every slot and user i's input stream being Bernoulli of rate $\lambda_i < \mu$, the total network delay W_i satisfies $E[W_i] = \frac{N-1-\lambda}{\mu-\lambda_i}$, with expected value $\frac{1}{\mu-\lambda_i}$, which represents the asymptotic growth in delay as the data rates are pushed towards the capacity boundary. Furthermore, for a fixed load value of $\frac{\lambda}{\mu}$ the delay is $O(n)$, growing linearly with the size of the network.

A transmitting node does not delete a packet from its local memory once it transmits it to a relay. The packets can reach the destination either through the source or the relay, despite the fact that a relay already has the packet. In addition, it is shown that algorithms which do not use redundancy cannot achieve an average delay of less than $O(n)$ [24]. This means the relay algorithm is optimal among the algorithms not using redundancy. This outcome determines that redundancy cannot increase capacity, but it can improve delay. Nevertheless, the delay cannot be bounded arbitrarily. In [24] it is also shown that no algorithm (with or without redundancy), which restricts packets to two-hop paths can provide an average delay better than $O(\sqrt{n})$. This is true even for the optimal policy, under which a source sends duplicate packets to new relays whenever possible. Then the time required to reach the destination T_n (time for the source or a relay to enter the cell the destination belongs to) has a limiting behavior of $e^{-d} \leq \lim_{n\to\infty} \frac{E[T_n]}{\sqrt{n}} \leq \frac{2}{1-e^{-d}}$. In fact, this bound cannot be overcome, even if multi-user reception is permitted, i.e allow a transmitted packet to be received by all users in the cell with the intended receiver.

4.3. Practical Schemes based on Cell Splitting and Redundancy

In this subsection, we consider schemes that address the capacity-delay tradeoff, aiming at realizing practical values for the transmission rate and delay in wireless ad hoc networks. The most notable feature of this protocol group is to split the deployment region into cells and capitalize on this spatial node segregation for efficient scheduling.

The delay bounds presented previously can be achieved by two different protocols. The first one is called *In-Cell Feedback Scheme with $\sqrt{n}$ Redundancy* [24]. In every cell with at least two users, a random sender and a random receiver are selected with uniform probability over all users in the cell. With equal probability, the sender operates in source-to-relay mode or relay-to-destination mode. When operation takes place according to the first mode, the sender transmits a packet's serial number (SN) on every opportunity until $\sqrt{n}$ copies of the packet have been delivered to distinct users, or until it transmits directly to the destination. Using sequence and request numbers for the senders and receivers/relays, packets are delivered in the destination in order. If all users receive data for their destinations according to a Poisson process of rate λ_i, the network can stably support rates $\lambda_i < \tilde{\mu}$ where $\tilde{\mu}$ is the reduced network throughput $\tilde{\mu} = \frac{\gamma_n(1-e^{-d})}{4(2+d)\sqrt{n}}$ and γ_n is a sequence that converges to 1 as n goes to infinity. Moreover, the average end-to-end delay satisfies $E[W_i] \leq \frac{1}{2} + \frac{1/\tilde{\mu}}{1-\rho_i}$. The In-Cell Feedback Scheme achieves data rates of $O(1/\sqrt{n})$ with a delay bound of $O(\sqrt{n})$.

A better delay bound can be achieved by the second scheme, hereafter termed by *Fair Packet Flooding Protocol* [24]. This approach is a flooding-based protocol that further degrades throughput, yielding in parallel a lower delay bound. In this case, the two-hop relay restriction is relaxed. At any timeslot, among all packets contained in at least one user of the cell that have never been received by some other user in the same cell, the packet that arrived earlier is chosen, and it is transmitted to all other users in the cell. If no such packet exists, the nodes remain idle. Ties are broken according to a fairness criterion, in the sense that packets of session i are given top priority every n timeslots. The oldest packet not been delivered to all users is selected to acquire network resources. The network under the Fair Flooding protocol is stable whenever $\sum_i \lambda_i < \frac{1}{E[T_n]}$ and the average delay satisfies $E[W] \leq \frac{1}{2} + \frac{E[T_n]}{1-\rho}$, where $E[T_n] \leq \frac{\log n(1+d/2)}{\log 2(1-e^{-d/2})} + 1 + \frac{2}{d}(1+\log(n/2))$ and $\rho = \sum_i \lambda_i E[T_n]$.

The two protocols presented suggest an important aspect of mobile ad hoc networks. There exists a delay-rate tradeoff of the form $\frac{\text{delay}}{\text{rate}} \geq O(n)$. This means that in a rate-delay diagram there exists a corresponding rate-delay curve (defined by the above relation) and the performance of every protocol falls on or near this curve. Furthermore, a necessary condition for any conceivable routing and scheduling protocol that stabilizes the network with input rates λ, while maintaining bounded average end-to-end delay $\bar{D}$, is given by $\frac{\bar{D}}{\lambda} \geq \frac{N-d}{4d}(1-\log 2)$, implying that if $d = O(1)$, then $\frac{\bar{D}}{\lambda} \geq O(n)$. The latter illustrates, that no scheduling and routing algorithm can simultaneously yield low delay and high throughput, revealing the inherent tradeoff once more.

Two different schemes for node mobility in ad hoc networks are presented and respective capacity-delay analysis is performed in [19]. These approaches follow the variable-size cell splitting paradigm proposed in [22]. In the first, the entire network is divided into cells of square shape, each of size $w(n)$. Thus, cell sizes scale linearly with the number of nodes. Then for $w(n) \geq \frac{2\log n}{n}$, each cell has at least one node with high probability (w.h.p.), acting as source, destination or relay. In the first scheme nodes are restricted to move within a single cell. In this layout, each packet traverses a mean number of hops of the order $\frac{\Theta(\bar{L})}{\sqrt{w(n)}}$, where $\bar{L}$ is the mean distance between two uniformly and independently chosen source-destination nodes in the network. It is proven that for the first scheme with $w(n) = \frac{k\log n}{n}$ and $k \geq 2$ in order to guarantee connectivity, the associated throughput is $T_1(n) = \Theta\left(\sqrt{\frac{\log n}{n}}\right)$ and $D_1(n) = \Theta(\sqrt{n})$, yielding a gain of $\Theta(\log n)$ in throughput compared to the case of a static random ad hoc network. Nevertheless, this comes at the cost of increased delay. By allowing nodes to execute a restricted mobility pattern, it is possible to obtain a throughput gain over the static network model paradigm.

In the second scheme the network is divided into a fixed number of cells, irrespectively of the network population. Each cell has area $1/l$ and there are l cells in total. Nodes are uniformly spread and the mobiles are restricted to move only within a single cell. Connectivity is guaranteed if $\frac{1}{l} > \frac{2\log n}{n}$. The average number of hops traversed by a packet becomes

$\frac{\Theta(\bar{L})}{\sqrt{l}}$. The corresponding throughput $T_2(n) = \frac{1}{\sqrt{l}}\Theta(1)$ and delay $D_2(n) = \Theta(\frac{n}{l})$. The delay in this case is smaller by a factor of l compared to the results of the Basic Relay model.

4.4. Critical Delay Schemes

In all the previous approaches different network settings have been used (mobility models, protocol models, etc.) yielding different forms of the capacity-delay tradeoff. Consequently, the latter depends not only on the nature of the mobility model but on the network arrangement model as well [25].

Using the notion of critical delay[3], two generic classes of mobility models are proposed, namely *hybrid random walk protocols* and *discrete random direction models*. The first exhibit a continuous range of critical delays in-between those of the independent and identically distributed mobility model and random walk mobility model, while the second exhibit a continuous range of critical delays in-between those of the random waypoint mobility model and the Brownian mobility model.

In the hybrid random walk protocols, the unit square network area is divided into $n^{2\beta}$ cells of area $1/n^{2\beta}$ and $n^{1-2\beta}$ subcells, each of area $1/n$ (β is a parameter of the model in the interval $[0,0.5]$). In the discrete random direction models, the area is divided into n^{2h} squares of area $1/n^{2h}$ each. Parameter $h \in [0,0.5]$. Node speed scales as $v_n = \Theta(1/\sqrt{n})$. The authors choose the mobility parameters of the previous models such that each node moves an average distance of $\Theta(1/\sqrt{n})$. They also make the critical assumption that only the source (i.e. not a relay node) can initiate a replication, noting also the difference between relaying and replication (in the latter multiple copies of a packet traverse the network). However, by definition, prohibiting relays from initiating a replication, does not prohibit multihop relaying as well. Under this assumption the critical delay for the hybrid random model with parameter $\beta > 0$ scales as $\Omega(n^{2\beta}/\log n)$. The respective quantity for the random direction model with parameter $h > 0$ scales as $\Omega(n^{h+1/2}/\log n)$. The 2-hop delay for the hybrid random walk protocol is obtained as $E[T] = \Theta(n^{2\beta}\log n) + \Theta(n^{2\beta})n^{-2\beta}$. For $\beta < 1/2$, $E[T] = \Theta(n)$ and for $\beta = 1/2$ we have $E[T] = \Theta(n\log n)$. For the discrete random direction model the 2-hop delay is $E[T] = \Theta(n^{h+1/2\log n}) + \Theta(n)$. Consequently, for $h = 1/2$, $E[T] = \Theta(n\log n)$ and for $0 \le h < 1/2$, $E[T] = \Theta(n)$.

Under the Brownian motion mobility model a more radical capacity-delay result has been obtained. In [26] a per-node throughput of $\Theta(1)$ with $\Omega(\log n/\sigma_n^2)$ delay was achieved by using a two-hop relaying scheme, allowing replication only by the sources and only once. Furthermore, it was shown that even if delay can be constrained to be $O(n^c/\sigma_n^2)$ for some constant $c < 0$, one can only achieve a per-node throughput of $\Theta(1/\sqrt{n})$, which is the same as in the case of static networks. This means the capacity-delay tradeoff under the Brownian motion regime degenerates, namely there is virtually no tradeoff between capacity and delay. However, such asymptotic behavior does not rule out the possibility of some limited tradeoff between delay and capacity. For instance, an achievable per-node throughput of $\Theta(1)$ can be obtained with a delay of $\Theta(\log n/\sigma_n^2)$.

By adopting the aforementioned framework, it can be deduced that the mobility models considered in the literature are in some sense extreme: they either exhibit the smallest

[3]Critical delay is the minimum average delay that must be tolerated in order to achieve a throughput of $\omega(1/\sqrt{n})$.

critical delays or the largest critical delays among all mobility models having roughly the same two-hop delay [25]. Thus, on one extreme, there is almost no delay-capacity tradeoff under the Brownian motion model and random walk model, and, on the other extreme, there is a smooth capacity-delay tradeoff for a wide range of delays under the random waypoint mobility model and independent and identically distributed mobility model. An interesting insight is that the critical delay is inversely proportional to the characteristic path length (i.e. the distance that a node travels without changing direction). Thus, from an application point of view, a scenario where the nodes move over long distances without changing directions (as in the random waypoint mobility model) is more desirable than a scenario where nodes change directions over short distances (as in the Brownian motion model).

4.5. Capacity-Delay Tradeoff in Hybrid Networks

The capacity of hybrid networks was obtained in the literature for a network model identical to that of a static homogeneous-node network. In [27], a routing algorithm for such networks is proposed with the goal of achieving close-to optimal capacity while keeping the delay small, thus addressing the inherent tradeoff between the two parameters in hybrid networks as well. The considered ad hoc network consists of n static nodes and z mobile nodes all lying in a disk of unit area (of radius $1/\sqrt{\pi}$). All nodes are initially uniformly and randomly spread. The mobiles follow the Uniform Mobility model, in which, nodes select randomly directions in $(0, 2\pi]$ and move towards the selected direction for a distance d, which is an exponentially distributed random variable with mean μ. Nodes are reflected at the boundary. In such model, node positions are independent of each other and the steady state node distribution is uniform. Sources and destination nodes are static, while mobiles act only as relays. For such a hybrid network in which n sender-receiver pairs are chosen randomly according to a uniform distribution among the static nodes, there exists a constant $c > 0$, such that each sender can achieve an average throughput of $c\frac{W\min(z,n)}{n\log^3 n}$, where W is the maximum available bandwidth. Moreover, the maximum delay incurred by the packet is at most $\frac{2d}{v}$, where d is the diameter of the network and v is the velocity of the mobile nodes. Note that the best achievable throughput by any algorithm with z mobile nodes is $O(Wz/n)$, while the aforementioned scheme provides a guarantee of $\Omega(Wz/(n\log^3 n))$.

The above throughput-delay bounds can be achieved by the routing protocol proposed in [27]. A static node S_1 wanting to send a message to destination R_1 transfers its message to a locally selected static leader node S, responsible for all communication in the neighborhood of S_1. S stores the message and waits for a mobile node M_1 such that M_1 is close enough to S and moving approximately along the direction of R. When such a node is available, S hands over the data to be transmitted to it. This is the *Static-to-Mobile phase*. The mobile nodes relay the packets intended for R_1 amongst themselves such that the packet moves closer and closer to the destination. This is the *Mobile-to-Mobile phase*. When the mobile relay carrying the packet is close enough to the destination, it hands off the packet to some leader node. This packet is then routed among the leader nodes towards the appropriate leader node, which then transmits the packet to the destination node. This is the *Static-to-Static* phase. Following this routing protocol, it is found that the delay experienced by a packet is inversely proportional to the speed of the mobile nodes and the throughput obtained is independent of the velocity of the mobile nodes. The protocol exploits the patterns in the

Table 1. Comparison of capacity-delay approaches

Mobility	Capacity	Delay	Model Features
static	$\Theta\left(\frac{1}{\sqrt{n\log n}}\right)$	$\Theta\left(\sqrt{\frac{n}{\log n}}\right)$	random node locations
random (i.i.d)	$\Theta(1)$	$\Theta\left(\frac{\sqrt{n}}{v(n)}\right)$	Basic Relay model
restricted-1D	$\Theta(1)$	$\Theta\left(\frac{\sqrt{n}}{v(n)}\right)$	great circles in unit-area sphere
cell-restricted (i.i.d)	$\frac{1}{\sqrt{\ell}}\Theta(1)$ $\Theta\left(\sqrt{\frac{\log n}{n}}\right)$	$\Theta\left(\frac{n}{\ell}\right)$ $\Theta(\sqrt{n})$	constant cell size $1/\ell$ cells varying as $\frac{\log n}{n}$
heterogeneous	$\Theta(\sqrt{n})$ -$\Theta(n)$	N/Av.	various mobilities
multiple hops	$\Theta\left(\frac{1}{\sqrt{nw(n)\log n}}\right)$	$O\left(\frac{1}{v(n)\sqrt{w(n)}}\right)$	$v(n) = o\left(\sqrt{\frac{\log n}{n}}\right)$, variable cell sizes $w(n)$
2-hop redundancy	$O\left(\frac{1}{\sqrt{n}}\right)$	$O(\sqrt{n})$	In-Cell Feedback Scheme
multihop redundancy	$O\left(\frac{1}{n\log n}\right)$	$O(\log n)$	Fair-Flooding Scheme
random (i.i.d)	$O\left(\frac{1}{\sqrt{n}}\right)$	$O(\sqrt{n})$	Hybrid networks
Brownian motion	$O\left(\frac{1}{\sqrt{n}}\right)$	$O(n^c/\sigma_n^2)$	$c < 0$ constant

mobility of nodes to provide guarantees on the delay. Moreover, the throughput achieved by the algorithm is only a poly-logarithmic factor off from the optimal one.

5. Comparative Evaluation

In the previous sections, various schemes have been presented and their basic features and principles with respect to the capacity-delay tradeoff in wireless multihop networks were exemplified. In this subsection we summarize the most interesting performance results, along with their corresponding system descriptions and mobility characteristics. Table 1 collects the most important capacity-delay results for the various network models described previously.

Before we proceed with the comparison of the presented schemes, we need to note the fundamental differences between relaying, replication and flooding. In relaying, a node that transmits a packet to a neighbor, deletes the packet from its own queue, relying on the next hop neighbor for the delivery of the packet. On the contrary in replication, a copy of the packet is transmitted and the node having the original packet continues to carry it, until one of the copies of the packet reaches its final destination. In the second case, delivery is guaranteed but as multiple copies of the same packet are present concurrently in the network, the effective traffic-carrying capacity of the latter is decreased. Flooding is a hybrid of the two schemes in which nodes relay the packets they receive by copying and transmitting them to all their neighbors (apart from the one they received the packet themselves). Flooding can be very efficient in terms of delivery and delay but the overhead increases exponentially affecting negatively network capacity [28]. In the presented methods, only relaying was considered and in some cases replication, however the latter was employed only by the original packet sources and only once, so that the network does not become crowded by multiple copies of the same packet.

The first two rows of Table 1 constitute the basic models for comparison purposes with regard to capacity, corresponding to static and random mobility networks respectively. Furthermore, the second one provides additional delay benchmark bounds for the rest of the models. A GI/GI/1-FCFS discipline for the queues at nodes was assumed in order to obtain such delay performance in the Basic Relay model and the deployment region was divided into cells of size $w(n) \geq 2\frac{\log n}{n}$. Node locations were initially randomly and uniformly selected. Consequently, these two models constitute schemes mainly based on the randomness of the network and respective protocols that exploit such randomness are proposed. An interesting remark is that these asymptotic capacity-delay results are preserved even if mobility is restricted in one dimension.

More complex approaches are the ones that embed virtual cells over the deployment region and define more complex relaying schemes. Similar approaches need to be taken both for hybrid networks (which consist of nodes with different capabilities) and networks with nodes of heterogeneous mobility capabilities (however of identical networking capabilities) to obtain effective capacity-delay bounds. With respect to the latter, delay results were not available, and therefore this absence is denoted by 'N/Av.' in Table 1. On the contrary, for hybrid networks both capacity and delay results are available, but since the objectives of the respective study are different, the delay result is a tight bound, rather than a scaling law and the capacity result is a lower bound.

A critical remark derived from Table 1 is that on one extreme, there is almost no delay-capacity tradeoff under the Brownian motion model and random walk model, and, on the other extreme, there is a smooth capacity-delay tradeoff for a wide range of delays under the random waypoint mobility model and independent and identically distributed mobility model. Such extremes need to be carefully taken into account in various analyses and system designs, as they can be critical for the inter-dependence of system parameters and overall performance. Furthermore, by an overall inspection of Table 1, the importance of the employed network as well as mobility model emerges eminently. Different models, yield different bounds. However, it becomes apparent as a general trend that mobility, in general, assists capacity (up to a certain degree), while multiple hops are in favor of delay. Depending on the application, correct design choices need to be taken according to the provided capacity-delay results.

6. Conclusions and Future Work

In this chapter, we studied and discussed the capacity-delay tradeoff in wireless multihop networks from a theoretical perspective and asymptotic behavior analysis. The emerging tradeoff is a consequence of the mobility property of mobile wireless networks and the lack of central infrastructure exhibited by such topologies.

Initially, a brief overview of the most frequent mobility models in wireless networks was provided and the basic features of them were summarized. In the sequel, the fundamental results on the capacity of static and mobile networks were presented and some capacity-approaching techniques that are mainly based on relaying were described and analyzed. Such mechanisms inherently affect the delay of the packet delivery processes, giving rise to the aforementioned capacity-delay tradeoff in wireless ad hoc networks. In addition, several schemes that tradeoff capacity in order to meet delay constraints and vice-versa were provided. A qualitative comparison among the various approaches included in this chapter was also provided, along with the basic characteristics of the system model considered by each approach.

The capacity-delay curves obtained in the considered framework can be very difficultly realized in practical applications. Especially the predicted capacity gains and delay guarantees cannot be accurately achieved, as their derivation involved several simplifying assumptions and their nature is by construction of asymptotic nature. Nevertheless, such results are rather indicative of the fundamental parameters that control the tradeoff and they could be accordingly utilized for designing efficient practical schemes. Despite the additional effort needed to make possible direct application of the aforementioned results, the emerging trends defined by such asymptotic relations may constitute a solid basis for further research, both theoretical and industrial, in order to develop accurate and practical schemes for controlling efficiently and effectively the identified tradeoff.

In addition, mobility has several implications for other layers of the protocol stack, most notably the application layer. In light of the capacity-delay tradeoff, mobility may be used to control node density by using scalar field distributions [29]. After their initial deployment nodes may be moved to desired locations, according to a desired node vector distribution. However, such movement can have an impact on the capacity of the network and the delay scaling. The same idea can also be exploited for providing fault-tolerance [30]. Nodes can be re-located to compensate for neighboring nodes that for some reason fail to properly operate (node or channel-specific reasons) and perform their networking functions. Capacity and delay have to be again analyzed, in order to obtain accurate guarantees on their scaling. Mobility can be employed for increasing/controlling coverage in wireless sensor networks [31]. The capacity bounds achieved in this case remain valid if the mobility model is continuous and belongs to one of the categories previously presented. However, delay concerns have to be taken into account, calling for additional studies and research.

Several open issues remain to be explored for such a fundamental functionality of wireless multihop networks such as mobility and its implications in the operation of a wireless multihop network. Among the most important ones, appears to be the development of a unifying framework that provides the most general case analysis of mobility, capacity and delay and covers generic scenarios of network/mobility models. Furthermore, a more thorough comparison among relaying, replication and flooding needs to be performed, in order to identify the impact of each general class of multihop packet delivery and obtain quantitative characterization of the appropriateness of each approach with respect to different network and application scenarios.

References

[1] J.-Y. Le Boudec, and M. Vojnovic, "Perfect Simulation and Stationarity of a Class of Mobility Models", in *Proc. of 24th Annual Conference of the IEEE Computer and Communications Society (INFOCOM)*, vol. 4, pp. 2743-2754, Mar. 2005.

[2] J. Broch, D. A. Maltz, D. B. Johnson, Y.-C. Hu, and J. Jetcheva, "A Performance Comparison of Multi-hop Wireless Ad Hoc Network Routing Protocols", in *Proc. of 4th Annual ACM/IEEE International Conference on Mobile Computing and Networking (MobiCom)*, pp. 85-97, Oct. 1998.

[3] X. Hong, M. Gerla, G. Pei, and C.-C. Chiang, "A Group Mobility Model for Ad Hoc Wireless Networks", in *Proc. of 2nd ACM International Symposium on Modeling, Analysis and Simulation of Wireless and Mobile Systems (MSWiM)*, pp. 53-60, Aug. 1999.

[4] E. M. Royer, P. M. Melliar-Smith, and L. E. Moser, "An Analysis of the Optimum Node Density for Ad Hoc Mobile Networks", in Proc. *IEEE International Conference on Communications (ICC)*, vol. 3, pp. 857-861, Jun. 2001.

[5] B. Liang, and Z. J. Haas, "Predictive Destance-Based Mobility Management for PCS Networks", in *Proc. of 18th Annual Joint Conference of the IEEE Computer and Communications Societies (INFOCOM)*, vol.3, pp. 1377-1384, Mar. 1999.

[6] H. Xie, and D. J. Goodman, "Mobility Models and Biased Sampling Problem", in *Proc. of 2nd IEEE International Conference on Universal Personal Communications (ICUPC)*, vol.2, pp. 803-807, Oct. 1993.

[7] H. Xie, S. Tabbane, and D. J. Goodman, "Dynamic Location Area Management and Performance Analysis", in *Proc. of 43rd IEEE Vehicular Technology Conference (VTC)*, pp. 536-539, May 1993.

[8] R. Thomas, H. Gilbert, and G. Mazziotto, "Influence of the Movement of the Mobile Station on the Performance of a Radio Cellular Network", in *Proc. of the 3rd Nordic Seminar*, paper 9.4, Sept. 1988.

[9] F. De Rango, P. Fazio, and S. Marano, "Adaptive Reservation in WLAN Networks under Smooth Random Mobility Model", in *Proc. of 62nd IEEE Vehicular Technology Conference (VTC-Fall)*, pp. 1264- 1268, Sept. 2005.

[10] W. Xi, X. Tan, and J. S. Baras, "Gibbs Sampler-Based Self-Organization of Autonomous Swarms", *Automatica*, pp. 1107-1119, Jul. 2006.

[11] J. Tian, J. Hahner, C. Becker, I. Stepanov, and K. Rothermel, "Graph-based Mobility Model for Mobile Ad Hoc Network Simulation", in *Proc. of 35th Annual Simulation Symposium, in cooperation with the IEEE Computer Society and ACM*, Apr. 2002.

[12] P. Johansson, T. Larsson, N. Hedman, B. Mielczarek, and M. Degermark, "Scenario-based Performance Analysis of Routing Protocols for Mobile Ad-Hoc Networks", in

Proc. of the 5th International Conference on Mobile Computing and Networking, pp. 195-206, 1999.

[13] J. Wu, and F. Dai, "Mobility Control and its Applications in Mobile Ad Hoc Networks", *IEEE Network Magazine*, vol.18, no. 4, pp. 30-35, Jul./Aug. 2004.

[14] P. Gupta, and P. R. Kumar, "The Capacity of Wireless Networks", *IEEE Trans. on Inf. Theory*, vol. 46, no. 3, pp. 388-404, Mar. 2000.

[15] M. Grossglauser, and D. N. C. Tse, "Mobility Increases the Capacity of Ad Hoc Wireless Netwokrs", in *Proc. of 20th Annual Joint Conference of the IEEE Computer and Communications Societies (INFOCOM)*, vol. 3, pp. 1360 - 1369, Apr. 2001.

[16] M. Grossglauser, and D. N. C. Tse, "Mobility Increases the Capacity of Ad Hoc Wireless Netwokrs", *IEEE Trans. on Netw.*, vol. 10, no. 4, pp. 477-486, Aug. 2002.

[17] S. Ali Jafar, "Too Much Mobility Limits the Capacity of Wireless Ad Hoc Networks", *IEEE Trans. Inf. Theory*, vol. 51, no. 11, pp.3954-3965, Nov. 2005.

[18] S. N. Diggavi, M. Grossglauser, and D. N. C. Tse, "Even One-Dimensional Mobility Increases Ad Hoc Wireless Capacity", in *Proc. of IEEE International Symposium on Information Theory (ISIT)*, Jun.-Jul. 2002.

[19] R. M. Moraes, H. R. Sadjadpour, and J. J. Garcia-Luna Aceves, "Mobility-Capacity-Delay Tradeoff in Wireless Ad Hoc Networks", *Elsevier Journal on Ad Hoc Networks*, vol. 4, no. 5, pp. 607-620, Sept. 2006.

[20] M. Garetto, P. Giaccone, and E. Leonardi, "On the Capacity of Ad Hoc Wireless Networks Under General Node Mobility", in *Proc. of 26th IEEE International Conference on Computer Communications. (INFOCOM)*, pp. 357-365, May 2007.

[21] B. Liu, Z. Liu and D. Towsley, "On the Capacity of Hybrid Wireless Networks", in *Proc. of 22nd Annual Joint Conference of the IEEE Computer and Communications Societies (INFOCOM)*, vol.2, pp. 1543- 1552, Mar.-Apr. 2003.

[22] A. El Gamal, J. Mammen, B. Prabhakar, and D. Shah, "Throughput Delay Trade-off in Wireless Networks", in *Proc. 23rd Annual Joint Conference of the IEEE Computer and Communications Societies (INFOCOM)*, pp. 464-475, Apr. 2004.

[23] J. Mammen, and D. Shah, "Throughput and Delay in Random Wireless Networks: 1-D Mobility is Just As Good As 2-D", in *Proc. of IEEE International Symposium on Information Theory (ISIT)*, pp. 437, Jul. 2004.

[24] M. J. Neely, and E. Modiano, "Capacity and Delay Tradeoffs for Ad Hoc Mobile Networks", *IEEE Trans. on Inf. Theory*, vol.51, no.6, pp. 1917-1937, Jun. 2005.

[25] G. Sharma, R. Mazumdar, and N. B. Shroff, "Delay and Capacity TRafe-Offs in Mobile Ad Hoc Networks: A Global Perspective", *IEEE Trans. on Netw.*, vol.15, no.5, Oct. 2007.

[26] Z. Lin, G. Sharma, R. R. Mazumdar, and N. B. Shroff, "Degenerate Delay-Capacity Tradeoffs in Ad Hoc Networks with Brownian Motion", *IEEE Trans. on Inf. Theory*, vol. 52, no. 6, pp. 2777 - 2784, Jun. 2006.

[27] N. Bansal, and Z. Liu, "Capacity, Delay and Mobility in Wireless Ad Hoc Networks", in *Proc. of 22nd Annual Joint Conference of the IEEE Computer and Communications Societies (INFOCOM)*, pp. 1553-1563, Mar.-Apr. 2003.

[28] A. O. Stauffer and V. C. Barbosa, "Probabilistic Heuristics for Disseminating Information in Networks", in *IEEE Trans. on Netw.*, pp. 425-435, vol.15, vo.2, April 2007.

[29] B. Zhang, and G. S. Sukhatme, "Controlling Sensor Density Using Mobility", in *Proc. of 2nd IEEE Workshop on Embedded Networked Sensors*, pp. 141-149, May 2005.

[30] P. Basu, and J. Redi, "Movement Control Algorithms for Realization of Fault-Tolerant Ad-Hoc Robot Networks", *IEEE Network Magazine*, vol.18, no. 4, pp. 36-44, Jul./Aug. 2004.

[31] B. Liu, P. Brass, O. Dousse, P. Nain, and D. Towsley, "Mobility Improves Coverage of Sensor Networks", in *Proc. of ACM MobiHoc*, pp. 300-308, May 2005.

In: Cluster Computing and Multi-Hop Network Research ISBN 978-1-60876-186-9
Editors: C. Jimenez and M. Ortego, pp. 163-177

Chapter 8

GEOMETRIC NEURAL COMPUTING FOR 2D CONTOUR AND 3D SURFACE RECONSTRUCTION

Jorge Rivera-Rovelo and Eduardo Bayro-Corrochano
Centro de Investigacion y de Estudios Avanzados del I.P.N.,
CINVESTAV, Unidad Guadalajara,
Department of Electrical Engineering and Computer Science,
Guadalajara, Mexico

Abstract

In this work we present an algorithm to approximate the surface of 2D or 3D objects combining concepts from geometric algebra and artificial neural networks. Our approach is based on the self-organized neural network called Growing Neural Gas (GNG), incorporating versors of the geometric algebra in its neural units; such versors are the transformations that will be determined during the training stage, and then applied to a point to approximate the surface of the object. We also incorporate the information given by the generalized gradient vector flow to select automatically the input patterns, and also in the learning stage in order to improve the performance of the net. Several examples using medical images are presented, as well as images of automated visual inspection. Comparisons using snakes and a version of GSOM developed to work incorporating versors and gradient information, confirm that our approach is very promising. As a second application, it is shown a kind of morphing or registration procedure, namely the algorithm can be used when transforming one model at time t_1 into other at time t_2. We include also examples applying the same procedure, now extended to models based on spheres.

1. Introduction

To approximate the contour or surface of an object is a task that can be carried out by different methods. If we want to preserve the topology of the data, a good choice is the use of self-organizing neural networks [4]. In this work we propose a very advanced algorithm using early vision preprocessing and self-organizing neural computing in terms of geometric algebra techniques. We believe that the early vision preprocessing together with self-organizing neurocomputing resembles in certain manner the geometric visual processing in biological creatures.

The proposed approach uses the *Generalized Gradient Vector Flow (GGVF)* [7] to guide the automatic selection of the input patterns, as well as to guide the learning process of the self-organized neural network GNG [5], to obtain a set of transformations M expressed in the conformal geometric algebra framework. In this framework rigid body transformations of geometric entities (like points, lines, planes, circles, spheres) are expressed in compact form as operators called *versors*, which are applied in a multiplicatively way to any entity of the conformal geometric algebra. Thus, training the network we obtain the transformation that can be applied to entities resulting in the definition of the object contour or shape. The experimental results show applications in medical image processing and visual inspection tasks, confirming that our approach is very promising.

2. Geometric Algebra

The Geometric Algebra [1, 2, 3] $G_{p,q,r}$ is constructed over the vector space $\mathcal{V}^{p,q,r}$, where p, q, r denote the signature of the algebra; if $p \neq 0$ and $q = r = 0$, the metric is Euclidean; if only $r = 0$, the metric is pseudo euclidean; if $p \neq 0, q \neq 0, r \neq 0$, the metric is degenerate. In this algebra, we have the *geometric product* which is defined as in (1) for two vectors a, b, and has two parts: the inner product $a \cdot b$ is the symmetric part, while the wedge product $a \wedge b$ is the antisymmetric part.

$$ab = a \cdot b + a \wedge b. \tag{1}$$

The dimension of $G_{n=p,q,r}$ is 2^n, and G_n is constructed by the application of the geometric product over the vector basis e_i.

$$e_i e_j = \begin{cases} 1 & for \quad i = j \in 1, \cdots, p \\ -1 & for \quad i = j \in p+1, \cdots, p+q \\ 0 & for \quad i = j \in p+q+1, \cdots, p+q+r \\ e_i \wedge e_j & for \quad i \neq j \end{cases}$$

This leads to a basis for the entire algebra: $\{1\}, \{e_i\}, \{e_i \wedge e_j\}, \{e_i \wedge e_j \wedge e_k\}, \ldots, \{e_1 \wedge e_2 \wedge \ldots \wedge e_n\}$. Any multivector can be expressed in terms of this basis. In the n-D space there are multivectors of grade 0 (scalars), grade 1 (vectors), grade 2 (bivectors), grade 3 (trivectors)... up to grade n. This results in a basis for G_n containing elements of different grade called *blades* (e.g. scalars, vectors, bivectors, trivectors, etc): $1, e_1 ... e_{12} ... e_{123} ... I$, which is called *basis blade*; where the element of maximum grade is the pseudoscalar $I = e_1 \wedge e_2 ... \wedge e_n$. A linear combination of basis blades, all of the same grade k, is called k-vector. The linear combination of such k-vectors is called *multivector*, and multivectors with certain characteristics represent different geometric objects (as points, lines, planes, circles, spheres, etc), depending on the GA where we are working in. Given a multivector M, if we are interested in extracting only the blades of a given grade, we write $< M >_r$ where r is the grade of the blades we want to extract (obtaining an homogeneous multivector M' or a r-vector).

2.1. Conformal Geometric Algebra

To work in Conformal Geometric Algebra (CGA) $G_{4,1,0}$ means to embed the Euclidean space in a higher dimensional space with two extra basis vectors which have particular meaning; in this way, we represent particular objects of the Euclidean space with subspaces of the conformal space. The vectors we add are e_+ and e_-, which square to $1, -1$, respectively. With these two vectors, we define the null vectors

$$e_0 = \frac{1}{2}(e_- - e_+); \quad e_\infty = (e_- + e_+), \tag{2}$$

interpreted as the origin and the point at infinity, respectively. From now and in the rest of the paper, points in the 3D-Euclidean space will be denoted in lowercase letters, while conformal points in uppercase letters; also the conformal entities will be expressed in the *Inner Product Null Space* (IPNS), and not in the *Outer Product Null Space* unless it is specified explicitly. To map a point $x \in \mathcal{V}^3$ to the conformal space in $G_{4,1}$, we use

$$X = x + \frac{1}{2}x^2 e_\infty + e_0. \tag{3}$$

As mentioned before, we can use CGA to represent particular objects of the 3D-Euclidean space; the spheres are specially interesting because they are the basic entities in CGA from which other entities are derived. Spheres with center in c and radius ρ are represented as

$$S = c + \frac{1}{2}(c^2 - \rho^2)e_\infty + e_0 \tag{4}$$

In fact, we can think in conformal points X as degenerated spheres of radius $\rho = 0$. Let be X_1, X_2 two conformal points. If we subtract X_2 from X_1, we obtain

$$X_1 - X_2 = (x_1 - x_2) + \frac{1}{2}(x_1^2 - x_2^2)e_\infty + e_0 \tag{5}$$

and if we square this result, we obtain

$$(X_1 - X_2)^2 = (x_1 - x_2)^2 \tag{6}$$

So, if we want a measure of the euclidean distance between the two points, we can apply (6). To obtain the 3D magnitude of a vector, we simply multiply its conformal representation X as follows

$$|x| = \sqrt{-2(X \cdot e_0)} \tag{7}$$

The dot product between two conformal vectors X_1 and X_2 results in $X_1 \cdot X_2 = (x_1 \cdot x_2) - \frac{1}{2}x_1^2 - \frac{1}{2}x_2^2$; therefore

$$x_1 \cdot x_2 = X_1 \cdot X_2 + \frac{1}{2}x_1^2 + \frac{1}{2}x_2^2 \tag{8}$$

Then, the angle θ between two vectors X_1 and X_2 in their conformal representation can be computed using (8) and (7)

$$\theta = \arccos\left(\frac{X_1 \cdot X_2 + \frac{1}{2}x_1^2 + \frac{1}{2}x_2^2}{|x_1||x_2|}\right) \tag{9}$$

By the other hand, to compute the perpendicular vector $x_3 \in \mathcal{E}^3$ to the vectors X_1 and X_2, we first compute the wedge product $A = X_1 \wedge X_2$, and then we take the coefficients of the components e_{12}, e_{23}, e_{31} (represented as $< A >_{e_{12}}, < A >_{e_{23}}, < A >_{e_{31}}$, which define the plane $\mathbf{b}$ dual to the vector, to finally obtain the perpendicular vector multiplying by $I_E = e_1 \wedge e_2 \wedge e_3$

$$x3 = I_E(\mathbf{b}) \quad = \quad I_E(< A >_{e_{12}} \wedge e_{12} + < A >_{e_{23}} \wedge e_{23} + < A >_{e_{31}} \wedge e_{31}) \qquad (10)$$

Note that the bivector $\mathbf{b} =< A >_{e_{12}} \wedge e_{12} + < A >_{e_{23}} \wedge e_{23} + < A >_{e_{31}} \wedge e_{31}$ can be used to build a rotor as explained further. Reader is encouraged to see the CGA representation of other entities consulting [3]. All of such entities and its transformations can be managed easily using the rigid body motion operators described further.

2.2. Rigid Body Motion

In GA there exist specific operators named *versors* to model rotations, translations and dilations, and are called rotors, translators and dilators respectively. In general, a versor G is a multivector which can be expressed as the geometric product of nonsingular vectors

$$G = \pm a_1 a_2 ... a_k \qquad (11)$$

In CGA, such operators are defined by (12), (13) and (14), being R the rotor, T the translator, and D_λ the dilator.

$$\begin{aligned} R &= e^{\frac{1}{2}\theta \mathbf{b}}, & (12) \\ T &= e^{-\frac{t e_\infty}{2}}, & (13) \\ D_\lambda &= e^{\frac{-\log(\lambda) \wedge E}{2}}, & (14) \end{aligned}$$

where $\mathbf{b}$ is the bivector dual to the rotation axis, θ is the rotation angle, $t \in \mathcal{E}^3$ is the translation vector, λ is the factor of dilation and $E = e \wedge e_0$.

Such operators are applied to any entity of any dimension by multiplying the entity by the operator from the left, and by the reverse of the operator from the right. Let be X_i any entity in CGA; then to rotate it we do $X_1' = R X_1 \tilde{R}$, while to translate it we do $X_2' = T X_2 \tilde{T}$, and to dilate we use $X_3' = D_\lambda X_3 \tilde{D}_\lambda$.

3. Determining the Shape of an Object

To determine the shape of an object, we can use a topographic mapping which uses selected points of interest along the contour of the object to fit a low dimensional map to the high dimensional manifold of such contour. This mapping is commonly achieved by using self-organized neural networks as Kohonen's Maps (SOM) or Neural Gas (NG) [6]; however, if we desire a better topology preservation, we should not specify the number of neurons of the network *a priori* (as specified for neurons in SOM or NG, together with its neighborhood relations), but allow the network to grow using an incremental training algorithm, as in the case of the Growing Neural Gas (GNG) [5]. In this work we follow the idea of growing neural networks and present an approach based on the GNG algorithm to determine the shape of objects by means of applying versors of the CGA, resulting in a model easy to handle in post processing stages; a scheme of our approach is shown in Fig. 1. The neural

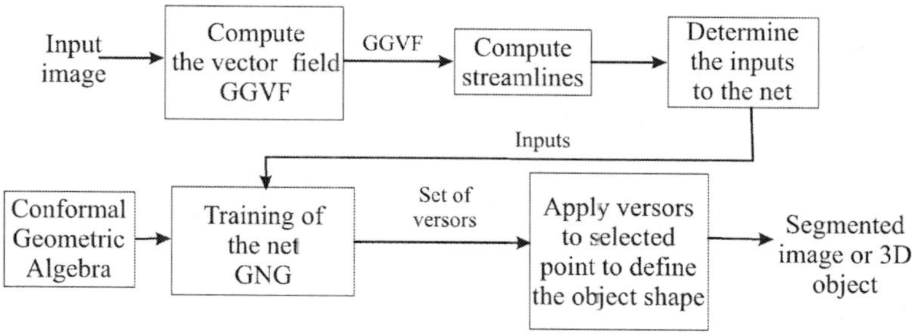

Figure 1. A block diagram of our approach.

network has versors associated to its neurons, and its learning algorithm determines their parameters that best fit the input patterns, allowing us to get every point on the contour by interpolation of such versors.

Additionally, we modify the acquisition of input patterns by adding a pre-processing stage which determines the inputs to the net; this is done by computing the Generalized Gradient Vector Flow (GGVF) and analyzing the *streamlines* followed by particles (points) placed on the vertices of small squares defined by dividing the 2D/3D space in such squares/cubes. The streamline or the path followed by a particle that is placed on $\mathbf{x} = (x, y, z)$ coordinates will be denoted as $S(\mathbf{x})$. The information obtained with GGVF is also used in the learning stage as explained further.

3.1. Automatic Samples Selection Using GGVF

In order to select automatically the input patterns, we use the GGVF [7], which is a *dense vector field* derived from the volumetric data by minimizing a certain energy functional in a variational framework. The minimization is achieved by solving linear partial differential equations which diffuses the gradient vectors computed from the volumetric data. To define the GGVF, the edge map is defined at first as

$$f(\mathbf{x}) : \Omega \rightarrow \mathcal{R} \tag{15}$$

For the 2D image, it is defined as $f(x, y) = -|\nabla G(x, y) * I(x, y)|^2$, where $I(x, y)$ is the gray level of the image on pixel (x, y), $G(x, y)$ is a 2D Gaussian function (for robustness in presence of noise), and ∇ is the gradient operator. With this edge map, the GGVF is defined as to be the vector field $\mathbf{v}(x, y, z) = [u(x, y, z), v(x, y, z), w(x, y, z)]$ that minimizes the energy functional

$$\mathcal{E} = \int\int g(|\nabla f|)\nabla^2 \mathbf{v} - h(|\nabla f|)(\mathbf{v} - \nabla f) \tag{16}$$

where

$$g(|\nabla f|) = e^{-\frac{|\nabla f|}{\mu}} \quad \text{and} \quad h(|\nabla f|) = 1 - g(|\nabla f|) \tag{17}$$

and μ is a coefficient. An example of such dense vector field obtained in a 2D image is shown in Fig. 2.a, while an example of the vector field for a volumetric data is shown in Fig. 2.b. Observe the large range of capture of the forces in the image. Due to this large capture range, if we put particles (points) on any place over the image, they can be guided

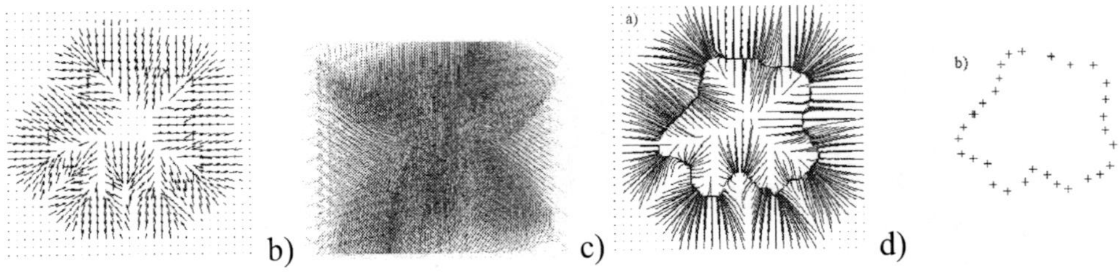

a) b) c) d)

Figure 2. Example of the dense vector field called GGVF (it is shown not all the vector field, but only representative samples of a grid). a) Samples of the vector field for a 2D image; b) Samples of the vector field for volumetric data. c) Example of streamlines for particles arranged in a 32x32 grid according to the vector field shown in (a); d) Points selected as input patterns according to (18).

to the contour of the object. The automatic selection of input patterns is done by analyzing the *streamlines* of points on a 3D grid topology defined over the volumetric data. It means that the algorithm follow the streamlines of each point of the grid, which will guide the point to the more evident contour of the object; then the algorithm selects the point where the streamline finds a peak in the edge map and gets its conformal representation X as in (3) to make the input pattern set. Additionally to the X (conformal position of the point), the inputs have the vector $\mathbf{v}_\zeta = [u,\ v,\ w]$ which is the value of the GGVF in such pixel and it will be used in the training stage as a parameter determining the amount of energy the input has to attract neurons. This information will be used in the training stage together with the position $\mathbf{x}$ for learning the topology of the data. Summarizing, the input set $\mathbf{I}$ will be

$$\mathbf{I} = \{\zeta_k = X_{\zeta_k}, \mathbf{v}_{\zeta_\mathbf{k}} | \mathbf{x}_\zeta \in S(\mathbf{x}') \text{ and } f(\mathbf{x}_\zeta) = 1\} \tag{18}$$

where X_ζ is the conformal representation of $\mathbf{x}_\zeta$; $\mathbf{x}_\zeta \in S(\mathbf{x}')$ means that $\mathbf{x}_\zeta$ is on the path followed by a particle placed in $\mathbf{x}'$, and $f(\mathbf{x}_\zeta)$ is the value of the edge map in position $\mathbf{x}_\zeta$ (assuming it is binarized). As some streamlines can carry to the same point or very close points, we can add constraints to avoid very close samples; one very simple restriction is that the candidate to be included in the input set must be at least at a fixed distance d_{thresh} of any other input.

Figure 2.c shows the streamlines according to the vector field shown in Fig. 2.a and the input patterns selected as described before are shown in Fig. 2.

3.2. Learning the Shape Using Versors

It is important to note that although we will explaining the algorithm using points, the versors can be applied to any entity in GA that we had selected to model the object. The network starts with a minimum number of versors (neural units) and new units are inserted successively. The network is specified by

- A set of units (neurons) named N, where each $\mathbf{n}_l \in N$ has its associated versor $M_{\mathbf{n}_l}$; each versor is the transformation that must be applied to a point to place it in the contour of the object. The set of transformations will ultimately describe the shape of the object.

- A set of connections between neurons defining the topological structure.

Also, take into account that

- There are two learning parameters: e_w and e_n, for the winner neuron and for the direct neighbors of it; such parameters remain constant during all the process.
- Each neuron $\mathbf{n_l}$ will be composed by its versor $M_{\mathbf{n_l}}$, the *signal counter* sc_l and the *relative signal frequency* rsf_l. The signal counter sc_l is incremented for the neuron $\mathbf{n_l}$ every time it is the winner neuron. The relative signal frequency rsf_l is defined as

$$rsf_l = \frac{sc_l}{\sum_{\forall \mathbf{n_j}} sc_j} \tag{19}$$

  This parameter will act as an indicator to insert new neural units.

With these elements, we define the learning algorithm of the GNG to find the versors that will define the contour as follows:

1. Let P_0 be a fixed initial point over which the transformations will be applied. This point corresponds to the conformal representation of p_0, which can be a random point or the centroid defined by the inputs. The initial transformations will be expressed as $M = e^{-\frac{t}{2}e_\infty}$ in the conformal geometric algebra. The vector t initially is a random displacement.
2. Start with the minimal number of neurons, which have associated random motors M as well as a vector $\mathbf{v_l} = [u_l, \ v_l, \ w_l]$ whose magnitude is interpreted as the capacity of learning for such neuron.
3. Select one input ζ from the inputs set $\mathbf{I}$ and find the winner neuron; that means to find the neuron n_l having the versor M_l which moves the point P_0 closer to such input:

$$M_{win} = \min_{\forall M} \sqrt{(X_\zeta - M P_0 \tilde{M})^2} \tag{20}$$

4. Modify M_{win} and all others versors of neighboring neurons M_l in such a way that the modified M will represent a transformation moving the point P_0 nearer the input. Note that each motor is composed by a rotation ΔR and a translation ΔT. The rotation is computed as in (12), being θ the angle between the actual position a and $a' = a + \mathbf{v}$ ($\mathbf{v}$ is the GGVF vector value in such position); the bivector dual to the rotation axis is computed as in (10); the rotors and translators are defined as

$$\Delta R_{win} = e^{\frac{e_w \, \eta(\mathbf{v}_\zeta, \mathbf{v_{win}})}{2}\theta \mathbf{b}} \tag{21}$$

$$\Delta T = e^{-\frac{\Delta t_{win}}{2}e_\infty} \tag{22}$$

$$\Delta t_{win} = e_w \, \eta(\mathbf{v}_\zeta, \mathbf{v_{win}}) \, (\mathbf{x}_\zeta - p_0) \tag{23}$$

   for the winner neuron, and

$$\Delta R_n = e^{\frac{e_n \, \eta(\mathbf{v}_\zeta, \mathbf{v_n})}{2}\theta \mathbf{b}} \tag{24}$$

$$\Delta T = e^{-\frac{\Delta t_n}{2}e_\infty} \tag{25}$$

$$\Delta t_n = e_n \, \eta(\mathbf{v}_\zeta, \mathbf{v_n}) \, (\mathbf{x}_\zeta - p_0) \tag{26}$$

for its direct neighbors, to obtain $\Delta M = \Delta T \Delta R$. Finally, the new motor is

$$M_{l_{new}} = \Delta M M_{l_{old}} \tag{27}$$

ϕ is a function defining the amount of a neuron can learn according to its distance to the winner one (defined as in (28)), and $\eta(\mathbf{v}_\zeta, \mathbf{v}_l)$ is defined as in (29).

$$\phi = e^{-\frac{(M_{win} P_0 \tilde{M}_{win} - M_l P_0 \tilde{M}_l)^2}{2\sigma}} \tag{28}$$

$$\eta(\mathbf{v}_\zeta, \mathbf{v}_l) = \|\mathbf{v}_\zeta - \mathbf{v}_l\|^2 \tag{29}$$

which is a function defining a quantity of learning depending on the strength to teach of the input ζ and the capacity to learn of the neuron, given in $\mathbf{v}_\zeta$ and $\mathbf{v}_l$, respectively. Also update

$$\mathbf{v}_{\mathbf{win}}^{\mathbf{new}} = [u_{win}^{new} \; v_{win}^{new} \; w_{win}^{new}]^T \tag{30}$$

$$\mathbf{v}_{\mathbf{n}}^{\mathbf{new}} = [u_n^{new} \; v_n^{new} \; w_n^{new}]^T \tag{31}$$

where

$$u_{win}^{new} = (u_{win} + e_w \, u_{win}) \quad v_{win}^{new} = (v_{win} + e_w \, v_{win}) \quad w_{win}^{new} = (w_{win} + e_w \, w_{win})$$

$$u_n^{new} = (u_n + e_n \, u_n) \quad v_n^{new} = (v_n + e_n \, v_n) \quad w_n^{new} = (w_n + e_n \, w_n)$$

5. Every certain number λ of iterations, determine the neuron with the rsf_l with highest value. Then, if any of the direct neighbors of that neuron is at a distance larger than c_{max}, do
   - Determine neighboring neurons $\mathbf{n_i}$ and $\mathbf{n_j}$
   - Create a new neuron n_{new} between $\mathbf{n_i}$ and $\mathbf{n_j}$ whose associated M and $\mathbf{v_l}$ will be

     $$M_{n_{new}} = \frac{M_i + M_j}{2}, \quad \mathbf{v_{l\_new}} = \frac{\mathbf{v_i} + \mathbf{v_j}}{2} \tag{32}$$

     the new units will have the values $sc_{new} = 0$ and $rsf_{new} = 0$.
   - Delete old edge connecting $\mathbf{n_i}$ and $\mathbf{n_j}$ and create two new edges connecting n_{new} with $\mathbf{n_i}$ and $\mathbf{n_j}$.
6. Repeat steps 3 to 5 if the stopping criterion is not achieved. The stop criterion is when a maximum number of neurons is reached, or when the learning capacity of neurons approaches to zero (is less that a threshold c_{min}), the first that happens will stop the learning process.

Training the network we find the set of M defining positions on a trajectory; such positions minimizes the error measured as the average distance between X_ζ and the result of $M_\zeta P_0 \tilde{M}_\zeta$:

$$\chi = \frac{\sum_{\forall \zeta}(\sqrt{(M_\zeta P_0 \tilde{M}_\zeta - X_\zeta)^2}}{N} \tag{33}$$

where M_ζ moves P_0 closer to input X_ζ, and N is the number of inputs.

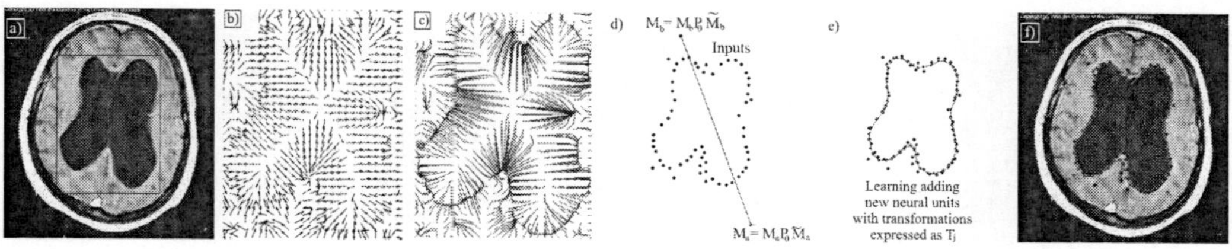

Figure 3. a) Original image and region of interest (ROI); b) Zoom of the dense vector field of the ROI; c) Zoom of the streamlines in ROI; d) Inputs and initial shape; e) Final shape defined according the 54 estimated motors; f) Image segmented according the results.

Figure 4. Application in visual inspection tasks: a) Original image and the region of interest (ROI); b) Zoom of the dense vector field of the ROI; c) Zoom of the streamlines in ROI; d) Inputs and initial shape according the two initial random transformations M_a and M_b; e) Final shape defined according the 15 estimated motors (original image with the segmented object).

4. Experiments

The Fig. 3 shows the result when algorithm is applied to a magnetic resonance image (MRI); the goal is to obtain the shape of the ventricle. Fig. 3.a shows the original brain image and the region of interest (ROI); Fig. 3.b the computed vector field for the ROI; Fig. 3.c the streamlines in the ROI defined for particles placed on the vertices of a 32x32 grid; Fig. 3.d shows the initial shape as defined for the two initial random motors M_a, M_b; Fig. 3.e shows the final shape obtained; and finally Fig. 3.f the original image with the segmented object.

Figure 4 shows an image showing that our approach can also be used for automated visual inspection tasks; reader can observe that such image contains a very blurred object. That image is for the inspection of hard disk head sliders. Figure 4.a shows the original image and the region of interest (ROI); Fig. 4.b the computed vector field of the ROI; Fig. 4.c the streamlines defined for particles placed on the vertices of a 32x32 grid; Fig. 4.d shows the inputs selected according the streamlines and the initial shape as defined for the two initial random motors M_a, M_b; Fig. 4.e shows the final shape obtained overlapped with the original image, showing that the algorithm gives good results if it is used for segmentation.

Figure 5 shows the application of the ggvf-snakes algorithm in the same problem. It is important to note that although the approaches of Fig. 4 and Fig. 5 use GGVF information to find the shape of an object, the estimated final shape is better using the neural approach than the one using active contours; the second approach (see Fig. 5) fails to segment the object no matter if the initialization of the snake is given inside, outside or over the contour

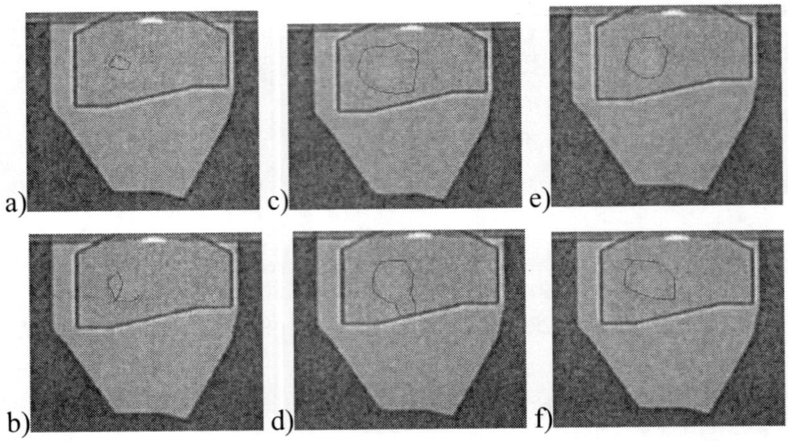

Figure 5. Result obtained when using the active contour approach to segment the object in the same image as in Fig. 4. a) Initialization of snake inside the object; b) Final result obtained with initialization showed in (a); c) Initialization of snake outside the object; d) Final result obtained with initialization showed in (c); e) Initialization of snake over the contour; f) Final result obtained with initialization showed in (e)

Table 1. Errors obtained by the algorithm with and without the GGVF information. ϵ_1: error without GGVF; ϵ_2: error with GGVF.

Example	ϵ_1	ϵ_2	Example	ϵ_1	ϵ_2
Ventricle 1	3.29	2.51	Eye 1	7.63	6.8
Eye 2	3.43	2.98	Column disk 1	4.65	4.1
Tumor 1	3.41	2.85	Tumor 2	2.95	2.41
Free form curve	2.84	1.97	Column disk 2	2.9	2.5

we are interested in. Additionally, the fact of expressing such shape as a set of motors allows us to have a model best suited to be used in further applications which can require the deformation of the model, specially if such model is not based on points but in other of the GA entities, because we do not need to change the motors (remember that they are applied in the same way to any other entity).

The proposed algorithm was applied to different sets of medical images. Fig. 6 show some images of such sets. The first row of each figure shows the original image and the region of interest, while the second row shows the result of the proposed approach. Table 1 shows the errors obtained with our approach using and not using the GGVF information. We can observe that the inclusion of GGVF information improves the approximation of the surface.

To compare our algorithm, we use the GNG with and without GGVF information, as well as a growing version of SOM, also using and not using the GGVF information. These algorithms were applied to a set of 2D medical images (some obtained with computer tomography (CT) and some with magnetic resonance (MR)). Figure 7.a the average errors when GSOM stops for different examples: segmenting a ventricle, a blurred object, a free

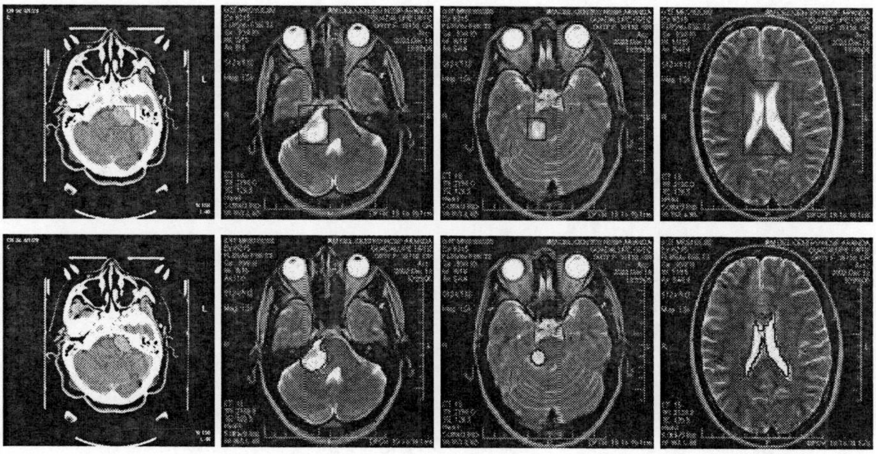

Figure 6. First row (upper row): original image and the region of interest. Second row: Result of segmentation

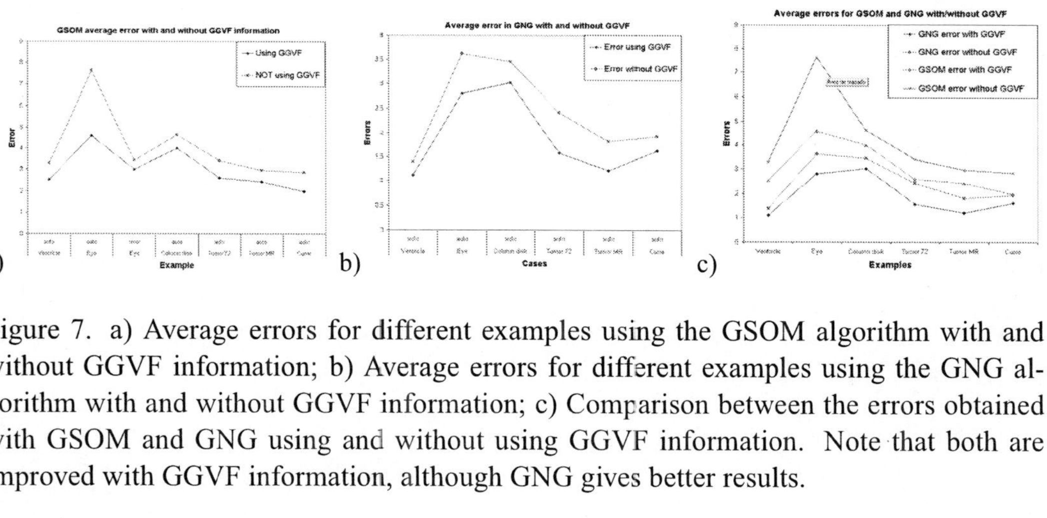

a) b) c)

Figure 7. a) Average errors for different examples using the GSOM algorithm with and without GGVF information; b) Average errors for different examples using the GNG algorithm with and without GGVF information; c) Comparison between the errors obtained with GSOM and GNG using and without using GGVF information. Note that both are improved with GGVF information, although GNG gives better results.

form curve, and a column disk. Note that using the GGVF information the error is reduced. That means that using the GGVF information, as we propose, it can be obtained a better approximation of the object shape. Figure 7.b shows the average errors obtained for several examples but using the GNG with and without the GGVF information. Note that again, the GGVF contributes to obtain a better approximation to the object surface. Also note that the average errors obtained with the GNG algorithm are smaller than the errors obtained with the GSOM, as can be seen in figure 7.c, and that both are improved with GGVF information, although GNG gives better results.

It is necessary to mention that the whole process is quick enough; in fact, the computational time required for all the images showed in this work, took only few seconds. The computation of the GGVF is the most time consuming task in the algorithm, but it only takes about 3 seconds for 64x64 images, 20 seconds for 256x256 images, and 110 seconds for 512x512 images. This is the reason why we decide do not compute it for the whole image, but for selected region of interest. The same criterion was applied to 3D examples.

Fig. 8.a shows the patient head with the tumor which surface we need to approximate; Fig. 8.b shows the vectors of the dense GGVF on a 3D grid arrangement of size 32x32x16; Fig. 8.c shows the inputs determined by GGVF and edge map, and also shows the initialization of the net GNG; Fig. 8.d-f show some stages of the adaptation process, while the net is determining the set of transformations M (Fig. 8.f is the final shape after training has finished with a total of 170 versors M (associated with 170 neural units)).

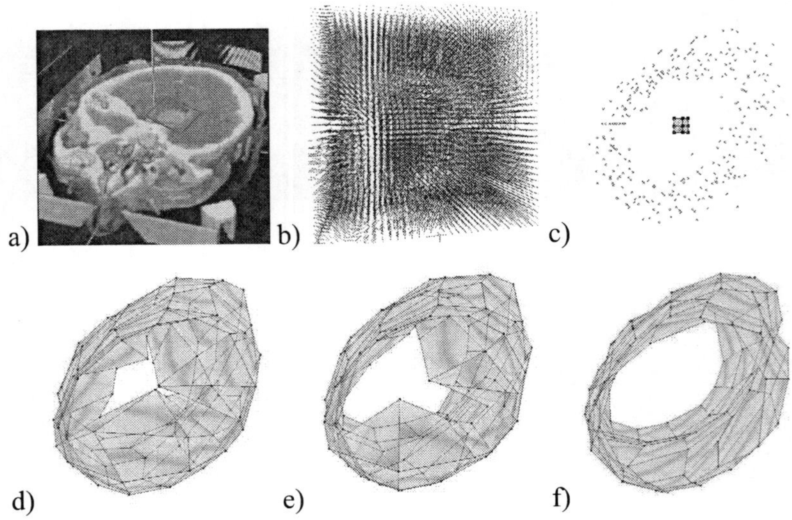

Figure 8. The algorithm for 3D object's shape determination. a) 3D model of the patient's head containing a tumor in the marked region; b) Vectors of the dense GGVF on a 3D grid arrangement of 32x32x16; c) Inputs determined by GGVF and edge map and the initialization of the net GNG; d-e) two stages during the learning; g) Final shape after training has finished with a total of 170 versors M (associated with 170 neural units).

Figure 9 shows other 3D example, corresponding to a pear, and the surface is well approximated. Figure 9.b shows the inputs and the initialization of the net with nine neural units (the topology of the net is defined as a sort of pyramid around the centroid of input points); while Fig. 9.c shows the result after the net has been reached the maximum number of neurons, which was fixed to 300; finally, Fig. 9.d shows the minimization of the error according to (33).

Another useful application of the algorithm using the gradient information of the GGVF during the training of the GNG neural net in the geometric algebra framework, is the transformation of one model obtained at time t_1 into another obtained at time t_2 (a kind of morphing of 3D surfaces).

Figure 10.a shows the initial shape which will be transformed into the one showed in 10.b; Fig. 10.c-f show some stages during the process. Note that Fig. 10.f looks like 10.b, as expected.

In the case shown in Fig. 11 we have one 3D model with an irregular shape which will be transformed to take a shape similar to a pear; Fig. 11.a shows the initial shape which will be transformed into the one showed in 11.b; Fig. 11.c-f show some stages during the process. Again the resulting volume looks like the one expected (Fig. 11.b).

To illustrate the application of the presented algorithm in cases having models based on

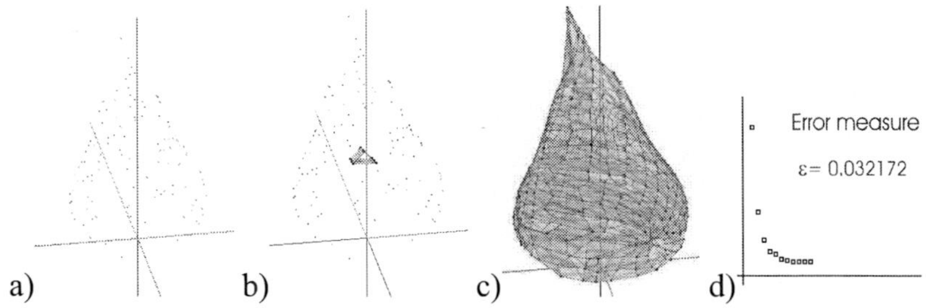

Figure 9. 3D object shape definition for the case of a pear. **a)** Inputs to the net selected using GGVF and streamlines; **b)** Inputs and the initialization of the net with nine neural units; **c)** Result after the net has been reached the maximum number of neurons (300 neurons); **d)** Error measurement using equation (33)

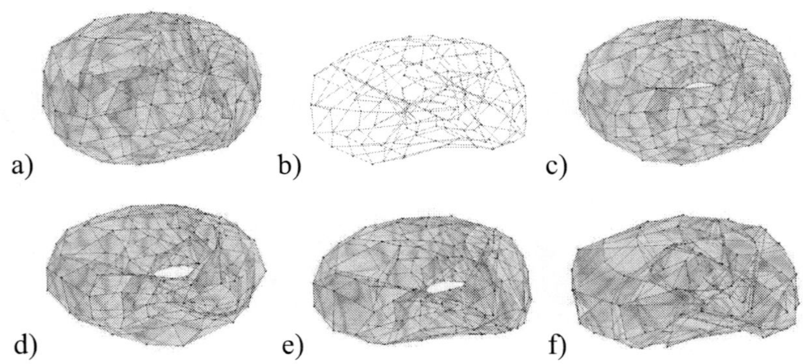

Figure 10. The 3D surface shown in (a) is transformed into the one shown in (b). Different stages during the evolution are shown in (c) to (f), where (f) is the final shape (that is, the final shape of a) after finishing the evolution of the net, which should look like (b).

entities different to the points, we show in Fig. 12 models based on spheres [8]. The goal is the same: morphing the model shown in Fig. 12.a and Fig. 12.d, to the one showed in Fig. 12.b and Fig. 12.e, respectively. The results are shown in Fig. 12.c and Fig. 12.f.

5. Conclusion

In this work the authors shown how to incorporate geometric algebra techniques in an artificial neural network approach to approximate 2D contours or 3D surfaces. In addition, they shown the use of the dense vector field named Generalized Gradient Vector Flow (GGVF) not only to select the inputs to the neural network GNG, but also as a parameter guiding its learning process. This network was used to find a set of transformations expressed in the conformal geometric algebra framework, which move a point by means of a versor along the contour of an object, defining by this way the shape of the object. This has the advantage that versors of the conformal geometric algebra can be used to transform any entity exactly in the same way: multiplying the entity from the left by M and from the right by $\widetilde{M}$.

There were presented some experiments showing the application of the proposed

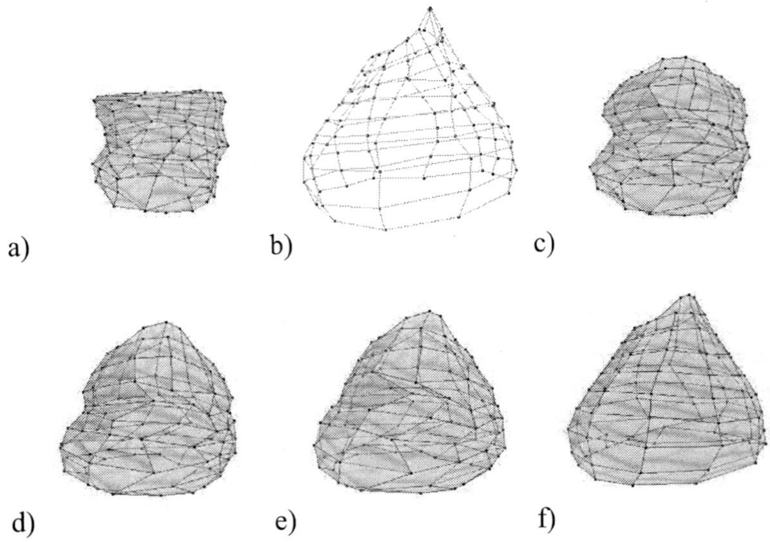

Figure 11. The 3D surface shown in (a) is transformed into the one shown in b). Different stages during the evolution are shown in (c) to (f), where (f) is the final shape (that is, the final shape of a) after finishing the evolution of the net, which should look like (b)

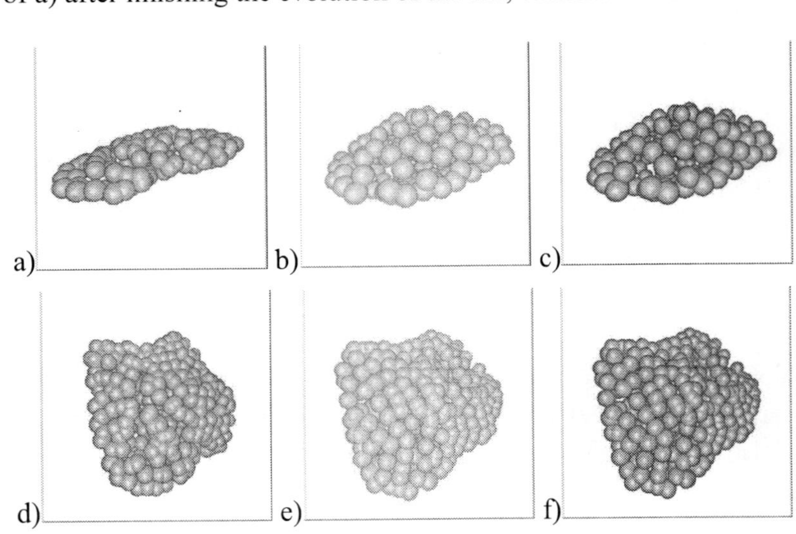

Figure 12. The 3D models based on spheres shown in (a) and (d) are transformed into the one shown in (b) and (e), respectively, resulting in the models showed in (c) and (f), respectively.

method in medical image processing and for visual inspection tasks. The results obtained show that by incorporating the GGVF information we can get automatically the set of inputs to the net, and also we improve its performance. Some comparisons between the results obtained with this algorithm, against the results obtained by a modified version of the GSOM net, and also against the ggvf-snakes, showed that our proposal is better. When dealing with the 3D case we presented two different applications: surface approximation and the transformation of a model at time t_1 onto other at time t_2, obtaining good results even using models based on spheres.

References

[1] E. Bayro-Corrochano, "Robot perception and action using conformal geometry", in *Handbook of Geometric Computing. Applications in Pattern Recognition, Computer Vision, Neurocomputing and Robotics*, E. Bayro-Corrochano (Ed.), Springer Verlag, Heidelberg, 2005, chap. 13, pp. 405-458.

[2] C. Perwass, and D. Hildenbrand, "Aspects of Geometric Algebra in Euclidean, Projective and Conformal Space", Christian-Albrechts-University of Kiel, Technical Report No. 0310, 2003.

[3] B. Rosenhahn, and G. Sommer, "Pose Estimation in Conformal Geometric Algebra", Christian-Albrechts-University of Kiel, Technical Report No. 0206, pp. 13-36, 2002.

[4] A. Angelopoulou, A. Psarrou, J. García Rodríguez, and K. Revett, "Automatic Landmarking of 2D Medical Shapes Using the Growing Neural Gas Network", in *Proceedings of the International Conference on Computer Vision, ICCV 2005*, October 13-21, Beijing, China, pp. 210-219.

[5] B. Fritzke, "A growing neural gas network learns topologies", *Advances in Neural Information Processing Systems* 7, MIT Press, Cambridge, MA, 1995.

[6] K. Mehrotra, C. Mohan, and S. Ranka, "Unsupervised learning", in *Elements of Artificial Neural Networks*, chap. 5, pp. 157-213.

[7] Ch. Xu, "Deformable models with applications to human cerebral cortex reconstruction from magnetic resonance images", Ph.D. Thesis, Johns Hopkins University, 1999, pp. 14-63.

[8] V. Ranjan and A. Fournier, "Union of spheres (UoS) model for volumetric data,", in *Proceedings of the Eleventh Annual Symposium on Computational Geometry*, Vancouver, Canada, 1995, C2-C3, pp. 402-403.

In: Cluster Computing and Multi-Hop Network Research ISBN 978-1-60876-186-9
Editors: C. Jimenez and M. Ortego, pp. 179-194

Chapter 9

MULTIVARIATE ZONATION OF LOGGING DATA

Danilo R. Velis*
Facultad de Ciencias Astronómicas y Geofísicas,
Universidad Nacional de La Plata, and CONICET, Argentina.

Abstract

The detection of stationary segments or zones in logging data sequences can be carried out by searching for change points in the data. These change points, which are associated to abrupt changes in the statistical properties of the underlying process, can be automatically identified by analyzing how the probability density function (pdf) of the data calculated in a short moving window varies with depth. Specifically, a change point is proposed when the difference between two consecutive pdf (or other statistical property) is significantly large. In a practical context, it is the statistical behavior of logging data (density, resistivity, porosity, etc.) that drives the zonation of a logging sequence into relatively homogeneous and stationary intervals, which are then associated to zones of similar lithologic character such as stratigraphic units and formations. When several logging data sequences are available, the segmentation can be carried out in a multivariate fashion. Here, all features of the individual sequences are captured simultaneously by the zonation algorithm. Naturally, this strategy leads to more reliable and consistent results because multiple data sequences are combined to produce a single sequence of segments. When applied to data from the Integrated Ocean Drilling Program (IODP), we show that the results of the multivariate zonation yields a high correlation with the available core-log lithology interpretation.

1. Introduction

The automatic or semiautomatic identification of zones or intervals in logging data measurements which are statistically distinct is an important data mining process. The zonation of a borehole environment is an essential step in the correlation of subsurface layers between wells, with application in oil exploration and reservoir evaluation. The problem can be viewed as a detection or an approximation problem. In the first case, a number of change

*E-mail address: velis@fcaglp.unlp.edu.ar. Address: Paseo del Bosque s/n, (B1900FWA) La Plata, Argentina.

points that delimit locally stationary segments are to be detected. In this context, the zonation is the dividing of a sequence into relatively homogeneous and stationary intervals, such that each segment is statistically distinctive from the adjacent ones. In the second case, the objective is to block or filter the sequence of observations to get a simpler data representation based on piecewise constant functions. Both approaches assume that the subsurface can be viewed as a layered model and that each layer is associated by distinct geophysical logging measurement. One important application is in well log analysis where it is desired to automatically subdivide the logging data into relatively uniform segments representing zones of similar lithologic character (stratigraphic units and formations). Segment boundaries correspond to abrupt changes in the layering, and conform the limits of relatively stable periods or geological meaningful zones. These elementary units of similar properties can then be used as the basis for inferring correlations between wells.

There is a myriad of different mathematical strategies to solve the zonation problem. Classical methods search for points with rapid changes in the mean [15, 2] or in the variance [3, 5, 2]. Other approaches are based on the method of maximum likelihood estimation [8], neural networks [11], cluster analysis [4], spectral analysis [7], etc. The approximation of logging data by means of piecewise segments has been solved also by various authors. See for example [6] in the context of sparse deconvolution and the references therein.

In this chapter, the focus is put into two different novel methods; (1) the detection of abrupt changes to identify statistically independent segments (statistical zonation) [14]; and (2) an edge-preserving filtering based in the approximation of the data sequence by piecewise constant functions (edge-preserving smoothing). In both cases, a full multivariate analysis is performed by considering various log data simultaneously. For this purpose, all available borehole measurements that were collected at the same well are combined into a single sequence that shares all the main characteristics of each sequence.

The first strategy [14], searches for abrupt changes in the first four moments of the data (i.e. mean, variance, skewness and kurtosis). Essentially, an analysis window of length $2L - 1$ is moved along the sequence and the probability density functions (pdf) of the two adjacent half-windows of length L are compared. When a significant difference between the two pdfs is detected, a change point is proposed. Smooth pdfs are estimated using the maximum entropy method as described in [13], which guarantees robustness when dealing with short data sequences. This involves the use of robust skewness and kurtosis measures that are not based on the conventional third and fourth-order moments around the mean.

The other strategy is based on a simple edge-preserving smoothing (EPS) of the data. This technique has been recently applied for smoothing and reducing noise in 3D impedance cubes, while preserving both lateral and vertical edges [1], and for preconditioning AVO inversions [9]. Here the 1D edge-preserving operator is applied for the zonation of well logs. Basically, this approach replaces each point by the average value of the most homogeneous segment in the neighborhood of the point. Further, a simple criteria is proposed to decide the number of homogeneous segments contained in the data.

The two zonation strategies are illustrated by analyzing simulated and real data sequences derived from well-logs. The real data comprises various borehole measurements which are part of the IODP, Leg 197, Site 1203 [12]. Here, the lithology interpretation based on extensive core samples and logging data analysis was available, so it was possible to make a comparison between this interpretation and proposed zonation results. Results show that in general, there is a high correlation between the published core-log lithology and the segmentations generated by the proposed procedures.

2. The Multivariate Zonation Problem

In the univariate zonation problem, let $\boldsymbol{x} = (x_1, x_2, \cdots, x_N)$ be a sequence of borehole measurements at depths $z_1, z_2, \cdots z_N$. The objective of the zonation process is to subdivide this sequence into homogeneous shorter segments so that (1) each depth-interval exhibits significantly different statistical properties, or (2) the whole data sequence can be conveniently approximated by a piecewise constant function. The points that delimit these intervals will be referred to as change points. For illustrative purposes, Figure 1 shows a piecewise sequence consisting of five segments or zones separated by four change points. The main objective of the zonation is to detect the points where the noisy data exhibits abrupt changes.

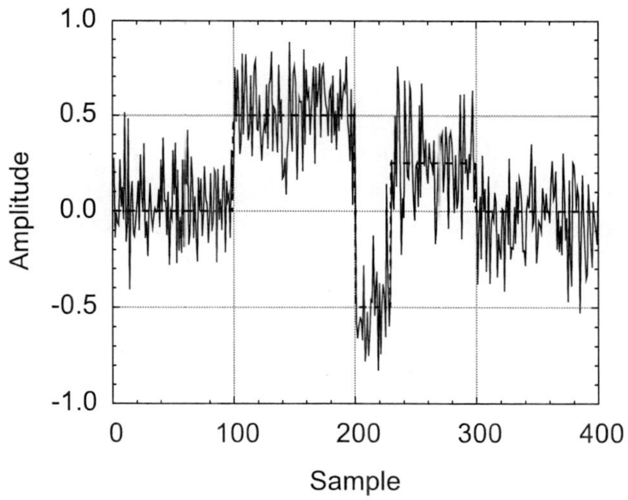

Figure 1. Piecewise constant function (dashed line) contaminated with random noise (solid black).

In this context, the zonation consists on searching for a sequence of M change points $(M \ll N)$

$$\zeta = (\zeta_1, \zeta_2, \cdots, \zeta_M), \qquad 1 < \zeta_1 < \zeta_2 < \cdots < \zeta_M < N, \tag{1}$$

which determine a set of $M + 1$ segments of length

$$\Delta_j = \zeta_{j+1} - \zeta_j + 1, \qquad j = 0, \cdots M, \tag{2}$$

where $\zeta_0 = 1$ and $\zeta_{M+1} = N$. The selection of the number of intervals M is a critical stage. The strategy I follow consists on gradually increasing M, selecting the best zonation for every value of M. When the number of segments does not significantly improve the partition, the process is terminated.

In the multivariate zonation case, one approach consists on segmenting each data sequence independently. This may lead to a different partition for each individual log, and will not be considered. Alternatively, I combine all well log data into a single sequence that exhibits the main features of all data. For this purpose, I define the "average" variable

$$\boldsymbol{y} = \frac{1}{K}\sum_{i=1}^{K} w_i \boldsymbol{x}_i^s, \tag{3}$$

where $\boldsymbol{x}_i^s$ is the standardized log sequence $\boldsymbol{x}_i$, K is the total number of log sequences involved, and w_i are weights.

2.1. Statistical Zonation

The first algorithm assumes that, given a change point ζ_j, the two adjacent segments have probability density functions which are significantly different [14]. In practice, an analysis window of length $2L-1$ is centered at a proposed change point ζ_j, and the pdfs of the data in the half-windows of length L on each side of the point j are estimated and compared. When they are significantly different, a change point is proposed. The algorithm proceeds adding one change point at a time.

In order to allow for the identification of short stationary segments, L must not be too large. For this reason, robust pdf estimates are required. The maximum entropy (MaxEnt) method with moment constraints described in [13] produces smooth non-parametric pdfs which are consistent with the data. In practice, these pdfs are obtained after maximizing the entropy of the pdf subject to moment constraints, which are directly estimated from the data. The constraints are given by the sample mean, variance, skewness and kurtosis, which are introduced into the optimization problem through Lagrange multipliers [13]. But instead of using the classical skewness and kurtosis, which are difficult to estimate for small samples, I use robust skewness and kurtosis measures that allows one to obtain more reliable pdf estimates.

I use a sliding window strategy to carry out the zonation. Here, an analyzing window of length $2L-1$ is moved along the data sequence and centered at the i-th point, $i = L, \cdots, N-L+1$. Then, the pdf corresponding to each half-window of length L is estimated using the mentioned MaxEnt method. Whenever a significant difference between the pdfs is observed, a candidate change point is assigned to the current point. In fact, at the first stage of the process the discrepancies between all possible adjacent pdfs are calculated. In a second stage, the M largest pdf discrepancies are selected to be associated to change points. To avoid the assigning of change points which are too close, I found it more appropriate to look for a single change point at a time, with the constraint that the next change point must lye at least a predefined number of samples away of all previous change points. Optimum change points are added until the next change point that is added does not yield a significant difference between the adjacent pdf or until the length of all the segments are smaller than a given minimum interval.

In particular, let $\hat{\zeta}$ be the current estimate of the j-th change point. Let $\boldsymbol{u} = (y_{\hat{\zeta}-L+1}, y_{\hat{\zeta}-L+1}, \cdots, y_{\hat{\zeta}})$ and $\boldsymbol{v} = (y_{\hat{\zeta}}, y_{\hat{\zeta}+1}, \cdots, y_{\hat{\zeta}+L-1})$ be the two subsets of $\boldsymbol{y}$ spanned by the analyzing window, and $\hat{p}_u(\boldsymbol{u})$ and $\hat{p}_v(\boldsymbol{v})$ be the corresponding estimated pdfs. I measure the discrepancy between $\hat{p}_u$ and $\hat{p}_v$ by calculating the difference between their respective cumulative distribution functions (cdf), $\hat{P}_u$ and $\hat{P}_v$, using the Kuiper test (the cdf's are calculated numerically by integrating the pdf estimates). The Kuiper test is a variant of the well known Kolmogorov-Smirnov test [10] that quantifies the maximum deviation above and below of the two cdfs. Contrarily to the Kolmogorov-Smirnov test, the Kuiper test is as sensitive in the tails as at the median of the distributions. The Kuiper statistics is given by

$$V = \max_{y_a \le y \le y_b} (\hat{P}_u - \hat{P}_v) + \max_{y_a \le y \le y_b} (\hat{P}_v - \hat{P}_u), \tag{4}$$

where y_a and y_b are the minimum and maximum values of $\boldsymbol{y}$. The asymptotic distribution of V is [10]

$$P(V > V_{obs}) = Q\left[\left(\sqrt{L} + 0.155 + 0.24/\sqrt{L}\right) V\right], \tag{5}$$

where

$$Q(\lambda) = 2\sum_{i=1}^{\infty}(4i^2\lambda^2 - 1)e^{-2i^2\lambda^2}. \tag{6}$$

Thus, the two distributions are significantly different when the probability P is small. As a consequence, for the zonation problem, candidate change points will be discarded based on their significance levels: a large value of P means that both $\boldsymbol{u}$ and $\boldsymbol{v}$ come from the same distribution (i.e. no change point), and vice-versa.

In summary, the statistical zonation algorithm can be described as a three stage process:

Stage 1:

Probability (5) is pre-calculated for every point in the range $(L, N - L + 1)$. This process implies the estimation of $2(N - 2L + 2)$ MaxEnt pdfs and their associated cdfs.

Stage 2:

One at a time, change points candidates are added according to the following strategy: at the beginning ($M = 1$), the point with the smallest probability is selected as a candidate for the first change point, yielding $\hat{\zeta}_1$ and the partition $(\zeta_0, \hat{\zeta}_1, \zeta_{M+1})$, which is comprised of two segments of lengths $\Delta_1 = \hat{\zeta}_1 - \zeta_0 + 1$ and $\Delta_2 = \zeta_{M+1} - \hat{\zeta}_1 + 1$, respectively. Then ($M = 2$), a new change point is added by selecting the smallest probability within the current longest segment (largest Δ_j), giving rise to a new partition $(\zeta_0, \hat{\zeta}_1, \hat{\zeta}_2, \zeta_{M+1})$ after rearranging in ascending order all previous change points.

This process is repeated and new change points are added (within the longest segments obtained so far) until: (1) all segments are shorter than a given minimum length, Δ_{min}; or (2) no change point can be added due to the constraint that any two adjacent change points must be separated by a minimum distance Δ_{sep}.

Stage 3:

All change points whose associated probabilities are larger than a predefined threshold are discarded. Also, the change points with largest probabilities in excess of a predefined number of change points are deleted.

Step by step, the algorithm is as follow:

1. Set $j = 0$.
2. For every point i in the initial search range $(L, N - L + 1)$,
   (a) set $\boldsymbol{u} = (y_{i-L+1}, y_{i-L+1}, \cdots, y_i)$ and $\boldsymbol{v} = (y_i, y_{i+1}, \cdots, y_{i+L-1})$,
   (b) estimate $\hat{p}_u$ and $\hat{p}_v$ using the MaxEnt method, and $\hat{P}_u$ and $\hat{P}_v$ by numerical integration,
   (c) compute V and probability P.
3. Set $j = j + 1$.
4. Add the change point $\hat{\zeta}_j$ as the point with the smallest P within the current search range.
5. Sort, in ascending order, the current set of change points and update the partition $(\zeta_0, \hat{\zeta}_1, \cdots, \hat{\zeta}_j, \zeta_{j+1})$.
6. Compute segment lengths according to Equation (2).
7. Update the search range: $(\zeta_{max} + \Delta_{sep}, \zeta_{max+1} - \Delta_{sep})$, where ζ_{max} is the beginning of the longest segment, $\Delta_{max} = \max_j \Delta_j$.
8. If $\Delta_{max} > \Delta_{min}$ (i.e. the longest segment is longer than a given minimum length), go to step 3.
9. Discard all change points whose probabilities are larger than a predefined significance level.
10. Discard all change points in excess of a predefined maximum number of change points whose probabilities are largest.

To illustrate and check the consistency of the statistical zonation algorithm, I applied it to the simulated sequence shown in Figure 1. The sequence was generated by adding random noise to a piecewise constant function consisting of five segments. The statistical zonation consists on detecting the four change points located at samples 100, 200, 230 ad 300.

For the segmentation I set $L = 21$ and $\Delta_{min} = \Delta_{sep} = 20$. Figure 2a shows the probability P as computed in the stage 1 (steps 1–2) of the above algorithm. As explained, the lowest probabilities are associated to large discrepancies between adjacent pdfs, because the Kuiper statistics is large. The four largest peaks (smallest P) in the figure correspond exactly to the four original change points. Note that the probability for the change point at sample 300 is the highest of the four. This is explained by the fact that the last two segments in the original data sequence (Figure 1) are separated by a moderate jump. The largest jump is at sample 200, and, as a consequence, the associated probability is the smallest.

The output of stage 2 (steps 3–8) of the above algorithm yields a set of fifteen probabilities arranged in ascending order. These values are plotted in Figure 2b, and also marked in Figure 2b. For the change points five and above, the probability increases more rapidly. This suggests that the number of homogeneous zones of the data can be estimated by the inspection of the probability curve shown in Figure 2b. In this particular case, a significance level of about 0.1 is enough to discard change points five and above. This is the result of the statistical zonation after stage 3 of the algorithm.

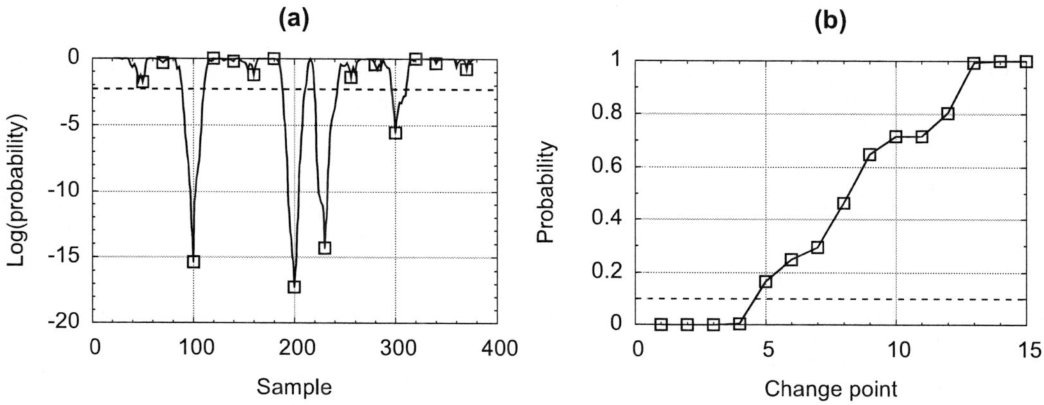

Figure 2. (a) Probability as computed after stage 1 of the statistical zonation algorithm. (b) Probabilities in ascending order. Change points five and above are not significant and hence discarded in the stage 3 of the algorithm.

2.2. Zonation by Edge-Preserving Smoothing

Edge-preserving smoothing (EPS) is a process aimed to filter noise while preserving abrupt changes (edges) in the data. The process can be generalized for two or more dimensions [1]. Figure 3 shows a step function contaminated with random noise filtered with a 91-point EPS operator. Clearly, all high frequency variability is removed, but the abrupt change at sample 100 is preserved.

The concept of the EPS is very simple, and can be summarized as follows. Let L be the length of the EPS operator. For every point i in the data sequence, $i = L, \cdots, N - L + 1$, the algorithm computes the means and the standard deviations of the L segments $\boldsymbol{u}_k = (y_{i-L+k}, \cdots, y_{i+1-k})$, $k = 1, \cdots, L$. Then, the mean corresponding to the segment with minimum standard deviation is assigned to the i-th output point location. Essentially, the filtering assigns the average value of the most homogeneous segment to the current point. When data exhibits high variability (i.e. when $\boldsymbol{u}_k$ includes the edge), the average value is not assigned, thus abrupt changes and edges are preserved, as shown in Figure 3.

In the zonation problem, EPS can be used to detect abrupt changes in the mean. The detection is carried out by identifying large variations in the first derivative of the filtered

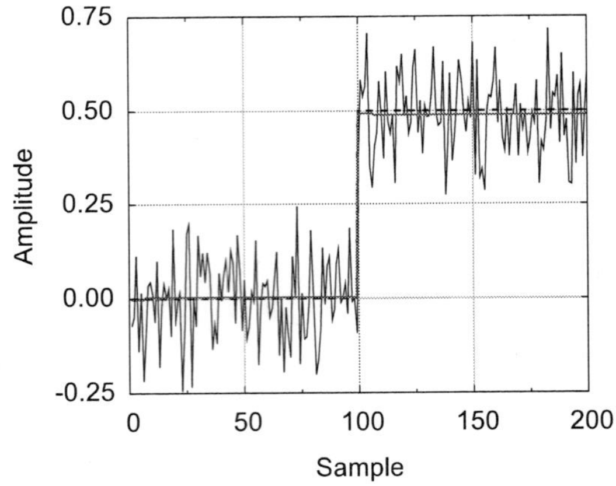

Figure 3. Step function (dashed line) contaminated with random noise (solid black). Red line shows the result after applying a 91-point EPS operator.

data. In Figure 3, for example, it is clear that the first derivative of the EPS output (red line) is largest at sample 100. Thus, the zonation problem reduces to the detection of abrupt changes in the filtered data.

In a practical context, the selection of the filter length is important because it controls the amount of smoothing and the number of change points that can be detected. If the filter is too short the smoothing may not be enough to adequately detect homogeneous segments or zones in the data[1]. On the other hand, if the filter is too long, some segments may not be detected, specially when two or more change points are very close. Figure 4 illustrates this concept. When a 41-point filter is used (red), the smoothing does not allow us to identify adequately the narrow zone between samples 200 and 230, because the filter is too long. But when a 21-point filter (blue) is utilized, the four abrupt changes at samples 100, 200, 230, and 300 are easily identified. In general, I found that the filter length must be in the order of the shortest segment that is expected to be detected.

The next step in the zonation requires a criterion to decide how many change points (or homogeneous zones) are needed to approximate the data using a set of segments of constant values. The criterion I use consists on selecting the least number of homogeneous segments that exhibit significantly distinct means. For this purpose, and as a second process of the zonation algorithm, the resulting segments are evaluated based on how significantly they contribute to the piecewise approximation. This strategy consists on selecting the M largest (positive or negative) first derivatives of the filtered data and assigning the average value to each of the resulting segments. In the example of Figure 4, there four largest jumps in the filtered data (blue line) lead to five homogeneous zones. More jumps (change points) could be selected, but these are much smaller than the first four. Then, the average of each zone is assigned to each segment and the piecewise approximation is obtained.

The goodness of fit of this approximation is measured by the root-mean-square error (RMSE), $E(M)$, between the original data sequence and the piecewise approximation:

[1]Note that for a filter of length unity, the EPS output is equal to the input data.

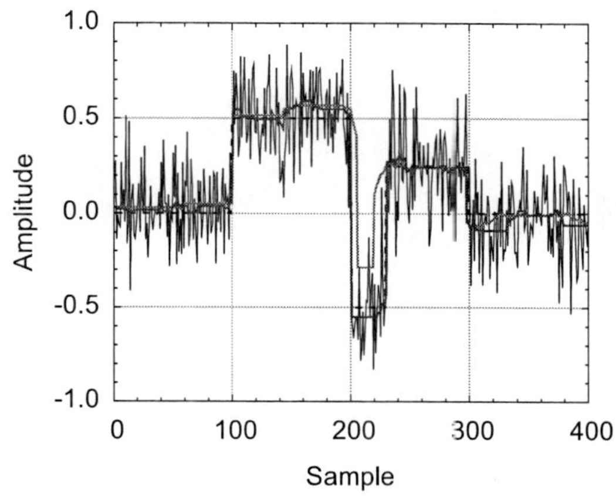

Figure 4. Piecewise constant function (dashed line) contaminated with random noise (solid black). Red and blue lines show the result after applying a 41 and a 21-point EPS operator, respectively.

$$E = \left[\frac{1}{N} \sum_{j=1}^{M+1} \sum_{i=\zeta_{j-1}}^{\zeta_j} (y_i - \bar{y}_j)^2 \right]^{1/2}, \tag{7}$$

where $\bar{y}_j$ is the average of the j-th segment:

$$\bar{y}_j = \frac{1}{\Delta_j} \sum_{i=\zeta_{j-1}}^{\zeta_j} y_i. \tag{8}$$

In general, as the number of change points increases, E decreases. For $M = 0$ (i.e. no change points), E is equal to the standard deviation of the data, since $(\zeta_0, \zeta_1) = (1, N)$ and $\Delta_1 = N$. For $M = N - 2$ (i.e. one change point at every location), $E = 0$, since $(\zeta_0, \zeta_1, \cdots, \zeta_{M+1}) = (1, 2, \cdots, N)$, $\Delta_j = 1$ and $\bar{y}_j = y_j$. In general, E leads to an L-curve as shown in Figure 5 for $M = 0, \cdots, 10$. Here, the change points were selected by keeping the M-largest (positive or negative) first derivatives of the EPS output (blue line in Figure 4). Clearly, for $M > 4$ the RMSE does not decrease significantly. This suggest that the number of change points of the data can be estimated by the inspection of the L-curve obtained when plotting E for an increasing number of change points. The final result of the zonation using four change points (five segments) is shown in Figure 6.

In summary, the zonation by edge-preserving smoothing involves three stages, likewise the statistical zonation,

Stage 1:

Select a filter length L, apply the EPS operator and calculate the absolute value of the derivative of the filtered sequence. The last step is used to detect the abrupt changes in the means. When the derivative is zero, there is no change point. When the derivative is large (positive or negative), there is a candidate change point.

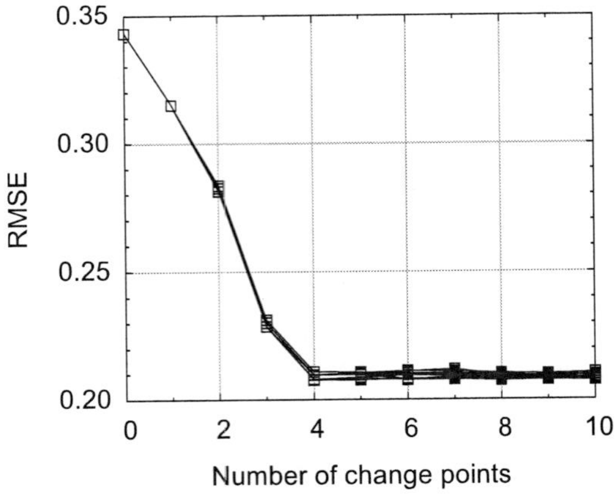

Figure 5. RMSE versus M for various EPS operator lengths in the range 11 to 25. The L-curves correspond to the example of Figure 4.

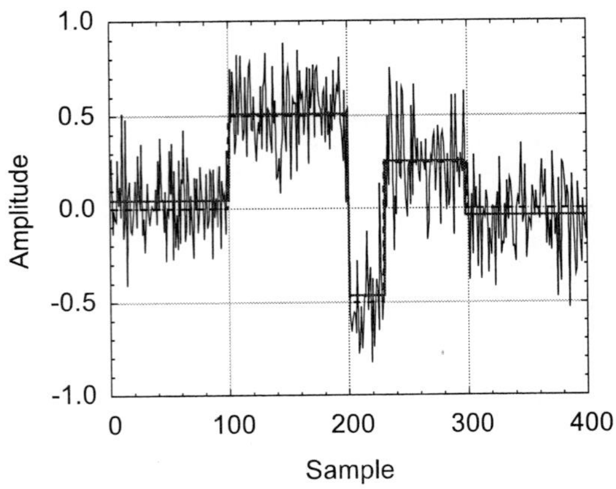

Figure 6. Final zonation (blue) using $M = 4$ and a 21-point EPS operator. Note that actually, any filter length in the range 11 to 25 produced the same segmentation.

Stage 2:

One at a time, change points candidates are added according to the following strategy: at the beginning ($M = 1$), the point with the largest derivative (positive or negative) is selected as a candidate for the first change point, yielding $\hat{\zeta}_1$ and the partition $(\zeta_0, \hat{\zeta}_1, \zeta_{M+1})$. Then ($M = 2$), a new change point is added by selecting the second largest derivative, giving rise to a new partition $(\zeta_0, \hat{\zeta}_1, \hat{\zeta}_2, \zeta_{M+1})$ after rearranging in ascending order all previous change points. This process is repeated and new change points are added until the derivatives are very small or a maximum number of candidate change points have already been added. Every time a new change point is added, the RMSE corresponding to that partition is computed.

Stage 3:

All change points that do not contribute significantly to the decrease of E are discarded. For this purpose, the L-curve E is analyzed to decide how many segments comprise the data (usually, the optimal number of change points corresponds to the corner of the L-curve). Also, the change points with the smallest contribution to the fit in excess of a predefined number of change points can be discarded if desired.

3. Results

The next example shows the results of the multivariate zonation when applied to various geophysical logging data sequences. The data, which are part of the Integrated Ocean Drilling Program (Leg 197, Site 1203), were collected to characterize the southward motion of the Hawaiian Hot-spot in the Emperor Seamount trend [12]. The drilling achieved moderate basement penetration and high recovery allowing for a detailed lithostratigraphy analysis. Downhole measurements, which are of very good quality in the basement sections, included various standard and non-standard tool strings and passes.

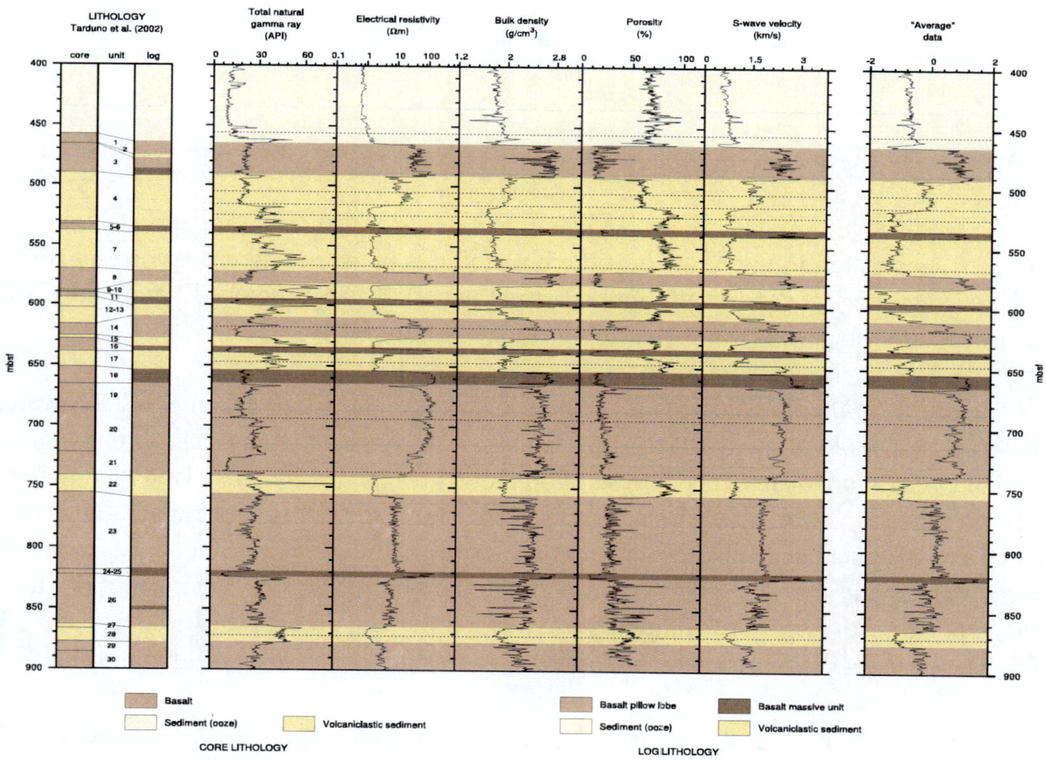

Figure 7. Statistical zonation of the logging data from the IODP (Leg 197, Site 1203). The left panel shows the core and log lithology columns after [12], together with the unit numbers. The center panel shows total natural gamma ray, electrical resistivity, bulk density, porosity and S-wave velocity. The right panel shows the "average" data actually used for the statistical zonation. The most significant change points (32 in total) are indicated either by colored bands (those that correlate well with the log lithology column) or dashed lines (those that do not correlate well with the log lithology column).

For the multivariate zonation, I selected a total of five borehole measurements: total natural gamma ray, electrical resistivity, bulk density, porosity and S-wave velocity. These data are shown in Figure 7 (center panel). The sampling interval is 0.1524 m, and each data sequence contains about 3300 samples in the considered interval. All data sequences were combined into a single log sequence according to equation (3), with $K = 5$ and weights $w_i = 1$ for electrical resistivity[2], bulk density and S-wave velocity, and $w_i = -1$ for natural gamma ray and porosity. This "average" sequence, which is plotted in the right panel of the figure, exhibits all main characteristics of the five individual log sequences. The zonation was carried out using the average sequence alone.

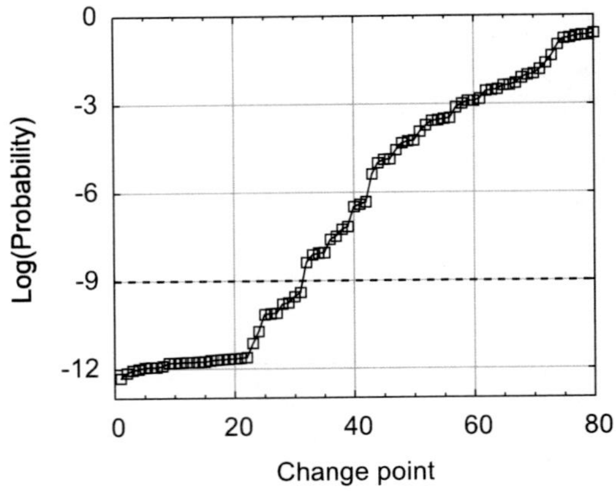

Figure 8. Logarithm to the base 10 of the probabilities in ascending order corresponding to the average log data shown in Figure 7 (right panel). Change points 32 and above are discarded in the stage 3 of the statistical zonation algorithm.

The results of the statistical zonation are shown in the same figure along with the available core-log lithology interpretation (left panel). The available interpretation in the analyzed interval (400–900 meters below sea-floor, mbsf) comprises 30 units of alternating sediments (ooze and volcaniclastic) and basalts [12]. The basement starts at about 460 mbsf. As for the statistical segmentation process we set $L = 31$, $\Delta_{sep} = 30$ and $\Delta_{min} = 50$ (sample units) and the 32 change points with the lowest probabilities were kept (all other change points with larger probabilities were discarded). The probabilities associated to all candidate change points are shown in Figure 8. In this particular case I set the threshold at $P = 10^{-9}$.

In general, the correlation between the statistical zonation and the core-log interpretation is from good to excellent. All the units identified in the available log interpretation were detected by the statistical zonation algorithm, except for the thin layers at about 475 mbsf (unit 2), 490 mbsf and 850 mbsf. It is worth mentioning that these thin layers are not easily identified by visual inspection of the logging data. Further, the last two layers were not identified in the core column at all. On the other hand, there is a total of 12 change points (denoted by dashed lines in the figure) which are not identified in the available log lithology interpretation. Some of these change points may be associated to units clearly identified in the core lithology. For example, units 27 and 28 are identified as a single unit

[2] In the actual computations, I used the logarithm of the electrical resistivity.

in the available log lithology, but as two units in the statistical zonation, in agreement with the core lithology analysis. The same can be said for units 19 and 20. Moreover, units 12, 13 and 14 are identified as two units in the available log interpretation, but correctly as three distinct units in the statistical segmentation. Finally, some of the remaining detected change points that do not correlate well with the available core-log interpretation may be associated to a fine lithology layering (e.g. division of units into sub-units, etc.). For example, the three change points detected in unit 4 by the statistical zonation at 505, 515 and 524 mbsf, correlate very well with core-lithology subunits 4i, 4k and 4m at 508, 516 and 523 mbsf, respectively. These subunits are not indicated in the core lithology column, but are detailed explained in [12]. A deeper analysis regarding subunits is beyond the scope of this chapter.

The results of the EPS zonation with a filter length $L = 25$ and $M = 20$ are shown in Figure 9a. The selection of the number of change points was done based on the analysis of the L-curves shown Figure 10, where the RMSE versus M is plotted for various filter lengths. Adding more than 20 change points does not contribute significantly to the piecewise approximation. Together with the EPS zonation, Figure 9a shows the 20 change points with lowest probabilities that were identified by the statistical zonation algorithm (red points). The correlation between the two results is excellent. All change points coincide almost exactly except for the change points at about 567 and 820 mbsf, only detected by one of the two zonation algorithms.

Figure 9b shows the results when $M = 32$ (same filter length). Though the twelve extra change points do not contribute significantly to the piecewise approximation (as shown in Figure 10), they permit to generate a finer partition of the logging sequence and to detect smaller zones of homogeneous data. For example, the region between 665 and 740 mbsf is subdivided into four zones by the EPS algorithm, and five zones by the statistical algorithm. Despite some minor differences, the overall correlation between the two zonation methods is very high (27 out of 32 change points coincide using both algorithms). As a consequence, the correlation between the EPS zonation and the available core-log lithology interpretations is also very high.

4. Conclusion

The detection of homogeneous segments or zones in logging data sequences can be carried out by searching for change points in the data. These points represent abrupt changes in the statistical nature of the data. The two proposed zonation algorithms allow one to identify semi-automatically a number of significantly distinct segments in the data. Multivariate zonation is obtained by combining various logging data sequences into a single "average" sequence that exhibits all the main characteristics of the individual components. This strategy guarantees that all the available information is taken into account and leads to a single partition of the whole data.

The statistical zonation method is based on the analysis of the probability density function of the data within a sliding window that is moved throughout the data sequence. When the discrepancy between the distributions on each side of the center of the window is significantly large, a change point is assigned. In this sense, the M change points with lowest probabilities (largest discrepancies) are kept, and those that exceed a predefined threshold are discarded. The threshold, which is used to set the number of segments in the data, can

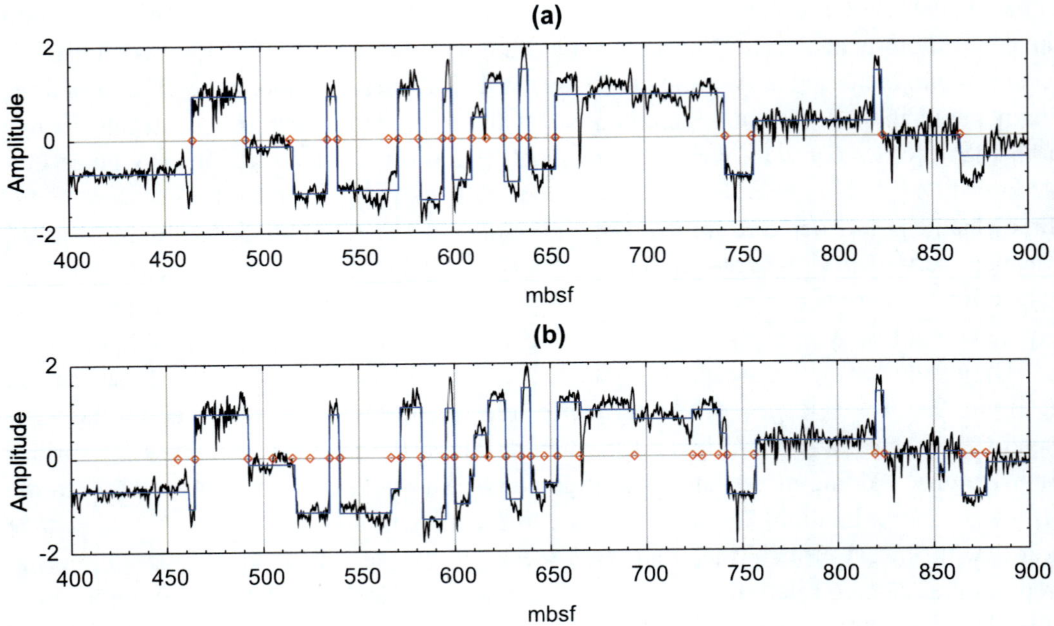

Figure 9. EPS zonation (blue) of the "average" logging sequence shown in the right panel of Figure 7. The derived piecewise approximations comprise a total of (a) 20 change points, and (b) 32 change points. For comparison, the red points show the 20 (a) and 32 (b) change points with lowest probabilities as obtained after the statistical zonation.

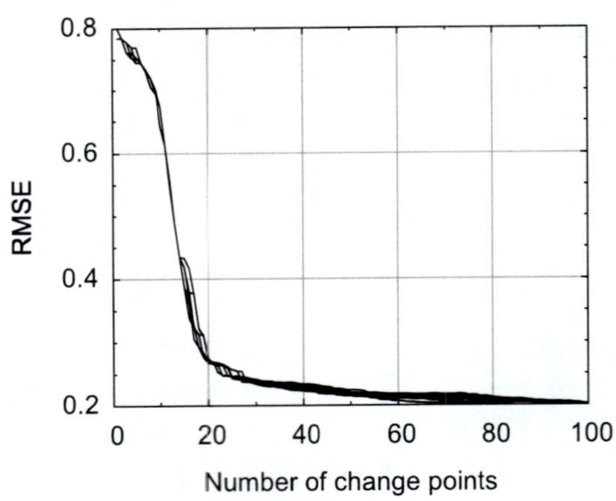

Figure 10. RMSE versus M for various EPS operator lengths in the range 15 to 30. The L-curves correspond to the "average" logging data shown in Figure 7.

be estimated by the analysis of the probability curve (probability that the two distributions are the same versus number of change points), which exhibits an abrupt increase for some M.

On the other hand, the edge-preserving smoothing (EPS) zonation is based on the assumption that the data can be represented by a piecewise approximation, which is in turn defined by a set of change points. After smoothing the data with an appropriate EPS operator, the main abrupt changes in the mean (edges) are detected by selecting the least number of homogeneous segments that yields an optimum fit to the data. For this purpose, the analysis of the L-curve (root-mean-square error of the fit versus number of change points) proved to be a good means to decide the number of segments in the data. Usually, the optimum zonation is obtained by selecting the number of change points at the point of maximum curvature (corner) of the L-curve.

So, knowledge of the number of segments is not required, but if available, both algorithms can be used to locate a given number of change points that lead to an optimum zonation. If the number of segments is overestimated, a finer segmentation is obtained. Nevertheless, since change points are added one at a time and the optimum segmentation is obtained in each case, it is possible to have a control over the number of segments which are significant for the zonation.

The results using synthetical data show that the number of change points and their exact locations can be estimated very accurately using any of the two techniques. Likewise, the results using logging data of the IODP (Leg 197, Site 1203) exhibit a high correlation with the available core-log lithology interpretation, where all main units and subunits were detected very accurately. Though the statistical zonation technique is more involved than the EPS technique in terms of data processing, both algorithms are computationally efficient, and long data sequences can be processed in less than a second using a personal computer.

The zonation algorithms presented in this chapter are viewed as auxiliary tools that may contribute useful information in the identification of the main lithology units and sequences as derived from measured geophysical log data. It is worth mentioning that the effectiveness of the methods are limited by the data quality and the assumption that different geologic settings (stratigraphic units and formations) can be associated to statistically different geophysical logging measurements. If there are no significant variations on the mean of the logging data, the EPS algorithms is prone to fail. The statistical algorithm, on the contrary, is more robust in the sense that it is based on the detection of changes in the probability density function, which is estimated from the mean, variance, skewness and kurtosis.

Acknowledgments

This work was partially supported by Consejo Nacional de Investigaciones Científicas y Técnicas, Argentina. This research used data provided by the Integrated Ocean Drilling Program.

References

[1] Nasher M. AlBinHassan, Yi Luo, and Mohammed N. Al-Faraj. 3D edge-preserving smoothing and applications. *Geophysics*, **71**(4):P5–P11, 2006.

[2] J. Davis. *Statistics and data analysis in Geology*. J. Wiley, New York, 1986.

[3] D. Gill. Application of a statistical zonation method to reservoir evaluation and digitized-log analysis. *American Association of Petroleum Geologists Bulletin*, **54**(5):719–729, 1970.

[4] D. Gill, A. Shomrony, and H. Fligelman. Numerical zonation of log suites and logfacies recognition by multivariate clustering. *American Association of Petroleum Geologists Bulletin*, **77**(10):1781–1791, 1993.

[5] D. M. Hawkins and D. F. Merriam. Optimal zonation of digitized sequential data. *Journal of Mathematical Geology*, **5**(4):389–395, 1973.

[6] Kjetil F. Kaaresen and Tofinn Taxt. Multichannel blind deconvolution of seismic signals. *Geophysics*, **63**(6):2093–2107, 1998.

[7] Uwe Ligges, Claus Weihs, and Petra Hasse-Becker. Detection of locally stationary segments in time series. In Wolfgang Härdle and Bernd Rönz, editors, *Proc. in Computational Statistics*, pages 285–290, Berlin, Germany, 2002.

[8] C.H. Mehta, S. Radhakrishnan, and G. Srikanth. Segmentation of well logs by maximum likelihood estimation. *Mathematical Geology*, **22**(7):853–869, 1990.

[9] Somanath Misra and Mauricio D. Sacchi. Global optimization with model-space preconditioning: Application to AVO inversion. *Geophysics*, **73**(5):R71–R82, 2008.

[10] W. H. Press, S.A. Teukolsky, W.T. Vetterling, and B.P. Flannery. *Numerical Recipes in FORTRAN: the Art of Scientific Computing, 2nd. edition.* Cambridge University Press, New York, 1992.

[11] S.J. Rogers, J.H. Fang, C.L. Karr, and D.A. Stanley. Determination of lithologies from well logs using neural networks. *American Association of Petroleum Geologists Bulletin*, **77**(5):731–739, 1992.

[12] J. A. Tarduno, R. A. Duncan, and D. W. Scholl. Motion of the Hawaiian hotspot: a paleomagnetic test, 2002.

[13] Danilo R. Velis. Estimating the distribution of primary reflection coefficients. *Geophysics*, **68**(4):1417–1422, 2003.

[14] D.R. Velis. Statistical segmentation of geophysical log data. *Mathematical Geology*, **39**(4):409–417, 2007.

[15] R. Webster. Automatic soil-boundary location from transect data. *Journal of Mathematical Geology*, **5**(1):27–37, 1973.

INDEX

C

D

E

F

K

L

M

Q

R

S

T

U

V

W

Y

Z